Human Factors in Engineering

This book addresses aspects of human factors in engineering and provides a detailed discussion of novel approaches, systems engineering tools, artificial cognitive systems, and intelligent technologies and automation. It presents applications in diverse areas, including digital manufacturing, transportation, infrastructure development, and cybersecurity.

This book:

- Merges the engineering perspective with the human factors and social dimension of computing and artificial intelligence–based technologies.
- Covers technological development of human factors engineering and the human dimension in applications across all areas of modern society.
- Relates to human behavior in the context of technology and systems interactions.
- Discusses the design and the appropriation of 3D printing techniques in the management of an innovative product system.
- Presents systems engineering tools, user experience methodologies, artificial cognitive systems, intelligent technologies, and automation.

The text is for students, professionals, and researchers in the fields of ergonomics, human factors, industrial engineering, and manufacturing engineering.

Human Factors in Design, Engineering, and Computing

Series Editors:
Tareq Ahram and Waldemar Karwowski

This series focuses on research and development efforts intended to promote the comprehensive integration of people and technological systems. To help foster integration of human factors research with emerging technologies, this series will focus on novel methodologies, design tools and solutions that advance our understanding of the nature of human factors and ergonomics with intelligent technologies and services. It will address all aspects of human factors and human-centered design, with a particular emphasis on applications of emerging technologies, computing, Artificial Intelligence and Systems. The series offers a multidisciplinary platform for researchers and practitioners alike, discussing emerging issues in the field of human factors engineering systems, with a special focus on (but not limited to) computing and AI-based technologies. It will be useful for researchers, senior graduate students, graduate students, and professionals in different domains including human factors, engineering design, systems science and engineering, and ergonomics.

Forthcoming titles:

Human Factors in Design
Intelligent Tools and Technological Innovations
Edited by Tareq Ahram and Waldemar Karwowski

Human Factors in Engineering
Manufacturing Systems, Automation, and Interactions
Edited by Beata Mrugalska, Tareq Ahram, and Waldemar Karwowski

Human Factors in Computing
Social Dimension and Artificial Intelligence-Based Technologies
Edited by Tareq Ahram and Waldemar Karwowski

Human Factors and Emerging Technologies
Integrating People and Technological Systems
Edited by Tareq Ahram and Waldemar Karwowski

Human Factors in Intelligence, Technology and Analytics
Edited by Tareq Ahram and Waldemar Karwowski

Human-Technology
Blending Artificial Intelligence, Computing, and Intelligent Design
Edited by Tareq Ahram and Waldemar Karwowski

For more information about this series, please visit: www.routledge.com/Human-Factors-In-Design-Engnineering-and-Computing/book-series/%20CRCHFIDEC

Human Factors in Engineering

Manufacturing Systems, Automation, and Interactions

Edited by
Beata Mrugalska, Tareq Ahram, and
Waldemar Karwowski

CRC Press
Taylor & Francis Group
Boca Raton London New York

CRC Press is an imprint of the
Taylor & Francis Group, an **informa** business

Designed cover image: © Shutterstock

First edition published 2023
by CRC Press
6000 Broken Sound Parkway NW, Suite 300, Boca Raton, FL 33487–2742

and by CRC Press
4 Park Square, Milton Park, Abingdon, Oxon, OX14 4RN

CRC Press is an imprint of Taylor & Francis Group, LLC

© 2023 selection and editorial matter, Beata Mrugalska, Tareq Ahram, and Waldemar Karwowski, individual chapters, the contributors

Library of Congress Cataloging-in-Publication Data
Names: Mrugalska, Beata, editor. | Ahram, Tareq Z., editor. | Karwowski, Waldemar,
    1953– editor.
Title: Human factors in engineering : manufacturing systems, automation, and
    interactions / edited by Beata Mrugalska, Tareq Ahram, and Waldemar Karwowski.
Description: First edition. | Boca Raton : CRC Press, [2023] | Series: Human factors in
    design, engineering, and computing | Includes bibliographical references and index.
Identifiers: LCCN 2022055089 (print) | LCCN 2022055090 (ebook) |
    ISBN 9781032370088 (hbk) | ISBN 9781032468235 (pbk) |
    ISBN 9781003383444 (ebk)
Subjects: LCSH: Human engineering.
Classification: LCC T59.7 .H8454 2023 (print) | LCC T59.7 (ebook) |
    DDC 620.8/2—dc23/eng/20230111
LC record available at https://lccn.loc.gov/2022055089
LC ebook record available at https://lccn.loc.gov/2022055090

ISBN: 978-1-032-37008-8 (hbk)
ISBN: 978-1-032-46823-5 (pbk)
ISBN: 978-1-003-38344-4 (ebk)

DOI: 10.1201/9781003383444

Typeset in Times
by Apex CoVantage, LLC

Contents

Preface

Competing in the global market is essential to business competitiveness in today's manufacturing industry. Manufacturers need to focus on innovations in product development, quality improvement, optimizing production schedules, reducing delivery time, and offering competitive value for consumers. It can be achieved by optimizing all elements of manufacturing processes, such as production methods, equipment, procedures, control, and information systems. Advanced information technologies that transfer human skills and manual activities to automated systems influence these processes and procedures. However, human operators are still important in manufacturing systems and should not be undermined or neglected.

Manufacturing companies must effectively manage the variety of complex factors associated with different aspects of advanced production systems and services. This book offers a unique perspective that blends the research from individual contributions presenting important domains in current manufacturing and production management applications. The covered topics include quality, health, and safety-oriented management models, human factors and ergonomics, analysis of organizational culture, as well as the consideration of the psychophysiological states of employees and work design. Further, we discuss the impact of strategic orientation and supply chain integration and provide many practical examples from diverse areas of applications. Special attention is given to the automotive industry, which constitutes the background for these studies. The competitive factors, productivity challenges, and the circular economy and lean manufacturing systems in the context of sustainability are discussed. Selected topics are devoted to developing system architectures, data analytics, and improving process and product quality. The modeling and simulation of epidemiological services are also presented. The last part of the book is devoted to quality improvement and process control of exemplary product parameters.

We believe that the strength of this book is embedded in its relatively intuitive and readable style, which illustrates the theory with practical examples. We hope this book will not only reach the students of production engineering, management, and applied psychology areas but also serve as a useful reference for researchers, practitioners, and industrial managers. We also hope to inspire others to focus their research on human aspects of contemporary production and service challenges.

We would like to express our gratitude to the authors who contributed to this book. Without their research and development efforts, this book would not have been possible.

Beata Mrugalska
Waldemar Karwowski
Tareq Ahram
Editors

October 2022

Editor Biographies

Beata Mrugalska, Ph.D., D.Sc., Eng., is Associate Professor and Head of Division of Applied Ergonomics in the Institute of Safety and Quality Engineering at the Faculty of Management Engineering at the Poznan University of Technology, Poznan, Poland. She received the M.Sc. Eng. in Management and Marketing (specialization: Corporate Management) in 2001 (including one semester under LPP Erasmus at Centria University of Applied Sciences in Ylivieska, Finland), the Ph.D. degree in Machines Building and Operations and D.Sc. (habilitation) degree in Mechanical Engineering in 2008 and 2019, all of them from the Poznan University of Technology, Poland. In 2012, she completed postgraduate studies in Occupational Health and Safety at the Faculty of Management Engineering at the Poznan University of Technology. She is the author or co-author/editor or co-editor of 10 books and more than 100 journal articles, book chapters and conference papers. She serves as a member of journal editorial board of *WORK: Journal of Prevention, Assessment & Rehabilitation, Human-Intelligent Systems Integration, International Journal of Occupational and Environmental Safety, International Journal of Ergonomics* (IJEG) and member of advisory board of *Journal of Turkish Operations Management.*

Tareq Ahram is presently working as an assistant research professor and lead scientist at the Institute for Advanced Systems Engineering at the University of Central Florida, USA. He received a Ph.D. degree in Industrial and Systems Engineering from the University of Central Florida, in 2008, with a specialization in human systems integration and large-scale information retrieval systems optimization. He has served as an invited speaker and a scientific member for several systems engineering, emerging technologies, neuro design, and human factors research, and as Invited Speaker and Program Committee Member for the U.S. Department of Defence Human Systems Integration and the Human Factors Engineering Technical Advisory Group. He serves as the Applications Subject editor member of *Frontiers in Neuroergonomics,* and the executive editor of the *Journal of Human-Intelligent Systems Integration.*

Waldemar Karwowski is currently working as Pegasus Professor and Chairman, Department of Industrial Engineering and Management Systems, and Executive Director, Institute for Advanced Systems Engineering, University of Central Florida, Orlando, USA. He holds a Ph.D. in Industrial Engineering from Texas Tech University. He is Past President of the International Ergonomics Association (2000–2003), and Past President of the Human Factors and Ergonomics Society (2007). He has over 500 publications focused on human performance, safety, cognitive systems engineering, human-centered design, neuro-fuzzy systems, nonlinear dynamics, neurotechnology, and neuroergonomics. He serves as Co-Editor-in-Chief of *Theoretical Issues in Ergonomics Science* journal (Taylor & Francis, Ltd), Editor-in-Chief of *Human-Intelligent Systems Integration* (Springer), and Field Chief Editor of *Frontiers in Neuroergonomics.*

Contributors

Junaid Ahmed
Shaheed Benazir Bhutto University
Sindh, Pakistan

Nehir Atila
Baskent University
Ankara, Turkey

Salvador Ávila Filho
UFBA—Federal University of Bahia
Salvador, Brazil

Manuel Ayala-Chauvin
Centro de Investigación en Ciencias
    Humanas y Educación (CICHE)
Universidad Tecnológica Indoamérica
Ambato, Ecuador

Marina Boronenko
Yugra State University
Khanty-Mansiysk, Russia Federation

Yuri Boronenko
Yugra State University
Khanty-Mansiysk, Russia Federation

Jorge Buele
SISAu Research Group, Facultad
    de Ingeniería y Tecnologías de la
    Información y Comunicación
Universidad Tecnológica Indoamérica
Ambato, Ecuador

Agata Chodowska-Wasilewska
Tecna Sp. Z o.o., Warsaw, Poland

Peng Lean Chong
Computer Engineering and Computer
    Science (CECS)
School of Engineering and Computing
    (SOEC), Manipal International
    University (MIU)
Negeri Sembilan, Malaysia

Atahan Erdağ
Baskent University
Ankara, Turkey

Iván Francisco Rodríguez-Gámez
Ciudad Juarez Autonomous University
Chihuahua, Mexico

Pavel Gulyaev
Yugra State University
Khanty-Mansiysk, Russia Federation

Juan Luis Hernández Arellano
Ciudad Juarez Autonomous
    University
Chihuahua, Mexico

Oksana Isaeva
Yugra State University
Khanty-Mansiysk, Russia Federation

Hyder Kamran
College of Business, University of Buraimi
Oman

M.G. Kanakana Katumba
Tshwane University of Technology
    (TUT), Pretoria Main Campus,
    South Africa, Faculty of
    Engineering and the Built
    Environment

Anton Kozhanov
State Scientific Center "Research
    Institute of Atomic Reactors",
    Dimitrovgrad, Russian Federation

Maria Kozhanova
State Scientific Center "Research
    Institute of Atomic Reactors",
    Dimitrovgrad, Russian
    Federation

Ernesto Lagarda Leyva
Sonora Institute of Technology
Sonora, Mexico

Opeyeolu Timothy Laseinde
The University of Johannesburg,
    Mechanical & Industrial Engineering
    Technology
South Africa

John M. Ikome
The University of Johannesburg,
    Mechanical & Industrial Engineering
    Technology
Johannesburg, South Africa

Aidé Aracely Maldonado-Macías
Ciudad Juarez Autonomous
    University
Chihuahua, Mexico

Jennifer Mayorga-Paguay
SISAu Research Group, Facultad
    de Ingeniería y Tecnologías de
    la Información y Comunicación,
    Universidad Tecnológica
    Indoamérica, Ambato,
    Ecuador

Beata Mrugalska
Poznan University of Technology
Poland

Robert Jeyakumar Nathan
Faculty of Economics
University of South Bohemia
    in Ceske Budejovice, České
    Budějovice, Czech Republic

Poh Kiat Ng
Faculty of Engineering and Technology,
    Multimedia University
Melaka, Malaysia

Yu Jin Ng
Department of Social Science and
    Humanities, Putrajaya Campus,
    College of Energy Economics and
    Social Sciences
Universiti Tenaga Nasional (UNITEN)
Kajang, Malaysia

Maciej Niemir
Łukasiewicz—Poznań Institute of
    Technology, Poznań University of
    Technology
Poland

Tadeusz Nowicki
Military University of Technology
Warsaw, Poland

Ibrahim Rashid Al Shamsi
College of Business, University of
    Buraimi, Oman

Yordán Rodríguez
National School of Public Health,
    Universidad de Antioquia
Colombia

İrem Sayın
Baskent University
Ankara, Turkey

F. Soner Alıcı
Baskent University
Ankara, Turkey

Angel Soria
Electrical and Computer Engineering,
    Purdue University
Lafayette USA

Yusuf Tansel İç
Baskent University
Ankara, Turkey

Sonia Umair
Institute of Business & Information
  Technology, University of the Punjab
Lahore, Pakistan

Umair Waqas
College of Business, University of
  Buraimi
Oman

Robert Waszkowski
Military University of Technology
Warsaw, Poland

Jian Ai Yeow
Faculty of Business
Multimedia University
Melaka, Malaysia

1 A New Ergonomics Management Model for Supply Chains

Iván Francisco Rodríguez-Gámez, Aidé Aracely Maldonado-Macías, Beata Mrugalska, Ernesto Lagarda Leyva, Juan Luis Hernández Arellano, and Yordán Rodríguez

CONTENTS

1.1 INTRODUCTION

Nowadays, management systems play a crucial role in the development of organizations as they allow them to fulfill the proposed objectives. Recently, more and more organizations have been implementing management systems in different fields such as quality (QMS) (Sfreddo et al., 2021; Rodríguez-Mantilla et al., 2020; Ingason, 2015; Psomas & Kafetzopoulos, 2014), environment (EMS) (Gunawan et al., 2020; Pacana & Ulewicz, 2017; Disterheft et al., 2012), health and safety (HSMS) (Morgado et al., 2019; Cąliş & Buÿükakinci, 2019a; Mohammadfam et al., 2017), and have emerged to generate integrated management systems (IMS). However, the latter have considered ergonomic aspects (Nunhes et al., 2019; Lima Marcos et al.,

2018; Domingues et al., 2016; Ifadiana & Soemirat, 2016; Yazdani et al., 2015a; Santos et al., 2013) although a proper management system for this discipline is lacking.

On the other hand, ergonomics management (EM) is used in the most recent research. However, it is usually related mainly to implementing ergonomics-related programs, even in IMS. However, in the absence of a clear and more accepted definition in the literature, it is difficult to conceive this term as a management system model proposed for quality, environmental, and even health and safety management systems (HSMS). Therefore, the term ergonomics management remains a "work-in-progress" concept. Even though modern approaches for quality management, as well as those for health and safety, have been clarifying some domains and characteristics through various models and standards, there is an opportunity for research considering them in the design of an ergonomics management model to extend its scope to the evaluation of entire supply chains (SC). In addition, the sustainable approach in supply chain ergonomics management from social sustainability (SS) perspective has received insufficient attention in both supply chain management (SCM) and supply chain sustainability issues addressed by researchers Korkulu and Bóna (2019) and Goethe et al. (2022). Although it is recognized that attempts have been made in the literature to establish a theoretical frame of reference for assessing SS through ergonomics that is mostly accepted for assessing social sustainability in supply chains, this objective has not yet been fully achieved. There are discrepancies and a lack of consensus among authors on its scope and conception (Simões et al., 2014). Additionally, Seuring (2012) states that social issues are little addressed in sustainable supply chain design, and some authors agree that there is a lack of research on sustainable supply chain management (SSCM) practices (Hong et al., 2018).

These findings evidence the need to consider the ergonomics approach for evaluation and improvement in the supply chain as a priority; so, researchers and stakeholders must have a better comprehension of the impact that ergonomics management has on employees' well-being and quality of life. In addition, the literature has recognized that the management and application of ergonomics generate economic benefits for those companies that have been successful in their implementation (Ciccarelli et al., 2022; Maldonado-Macías et al., 2021; Naeini et al., 2018; Sultan-Taïeb et al., 2017; Pereira Da Silva et al., 2012; Tompa et al., 2008). Therefore, an ergonomics management model must strongly emphasize the social sustainability of companies and SC. Accordingly, this chapter aims to conduct a systematic literature review of management systems and models to establish the basis for designing an ergonomics management model applicable to companies and to those SC in which they participate.

1.2 MAIN TOPIC LITERATURE

The International Organization for Standardization (ISO) defines a management system as a set of elements of an organization that are interrelated or interact to establish policies, objectives, and processes to meet the goals established by the organization. Such elements consider the structure of the organization, roles and responsibilities, planning, operation, performance evaluation, and continuous improvement. These are key elements of management that should be included in the design of any management system. The aforementioned literature review shows an overview of management systems, as well as standards that can be used in the conformation of the ergonomics management model. A quality management system (QMS) focuses on achieving the quality policy and quality objectives that drive to meet the company and customer requirements. The QMS is articulated through the facility integrity organization: its policies, procedures, and processes that are required to successfully achieve quality management of the facility (Deighton & Deighton, 2016).

On the other hand, HSMS refers to functions, processes, and tangible practices associated with occupational safety. According to ISO 45001:2018, an HSMS is to provide a framework for managing health and safety risks and opportunities. Their implementation is a strategic and operational decision for an organization. There are several HSMS-related standards available for companies to adapt and use. Regardless of the HSMS chosen, these systems are designed and based on a continuous improvement process to control hazards and risks to an acceptable level and improve worker health and safety (Labodová, 2004). One of the root causes of many industrial disasters is the absence of an HSMS (Bhasi et al., 2010; Haas & Yorio, 2021).

1.2.1 RELEVANT MANAGEMENT STANDARDS AND MODELS

In terms of standards and management models, two relevant ones are widely accepted by organizations to meet the needs of their customers and stakeholders through quality management. On the one hand, the implementation and certification of quality systems according to ISO 9000 is undoubtedly the most popular methodology. On the other hand, certification based on the European Foundation for Quality Management Model (EFQM) is gaining ground in improvement processes (Bayo-Moriones et al., 2011). By 2020, the ISO 9001:2015 standard was awarded 916,842 certifications worldwide (International Standard Organization, 2021). EFQM is currently used by more than 800 organizations across Europe (Gómez-López et al., 2019). Although both models focus on quality management, there are some important characteristics in their implementation approaches. Thus, ISO 9001:2008 promotes the adoption of a process approach model when developing, implementing, and improving the effectiveness of a quality management system to enhance customer satisfaction by meeting customer requirements (ISO 9001:2008, Quality Management Systems—Requirements). ISO 9000 standard proposes a process-based quality management system, as presented in Figure 1.1. Similarly, the EFQM 2020 model is based on process management as an accepted good practice since its objective is to help the sustained and sustainable growth of an organization through the continuous increase of value for all stakeholders by the efficient management of the transformation processes, plans, and projects. In addition, it has its management tool called RADAR (Result, Approach, Deploy, Assess, and Refine), whose function is to identify how the organization is working and what could be improved. The vision of the process approach makes it easier for organizations to adopt or manage any process, through identification, design, execution, measurement, and control. Considering these two main process-based models, when ergonomics is considered requirements or inputs must be processed and solutions or fulfillments are generated. Accordingly, ergonomics management can adopt some similarities of QMS to develop its proper model.

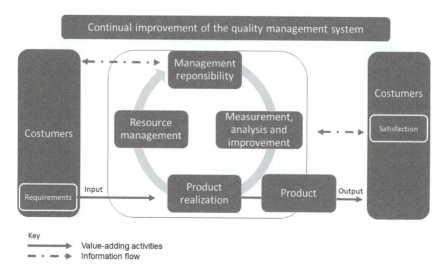

FIGURE 1.1 Process-based quality management system adapted from *ISO 9001:2008(En), Quality Management Systems—Requirements* (www.iso.org/obp/ui/#iso:std:iso:9001:ed-4:v2:en).

1.2.2 PLAN, DO, CHECK, ACT MODEL, APPLIED HEALTH, AND SAFETY STANDARDS

The PDCA cycle is a systematic process designed to obtain valuable learning and knowledge for the continuous improvement of a product, process, or service. It is used by QMS, EMS, HSMS, and Information Security Management Systems (ISMS), regulated by ISO, as well as in Total Quality Management (TQM) models. This management model is based on four phases aimed to develop continual improvement:

- *Plan:* Establish the objectives of the system and its processes and the resources needed to deliver results following customer's requirements and the organization's policies, and identify and address risks and opportunities.
- *Do:* Implement what was planned.
- *Check:* Monitor and (where applicable) measure processes and the resulting products and services against policies, objectives, requirements, and planned activities, and report the results.
- *Act:* Take actions to improve performance, as necessary.

This model has been used for versions of ANSI/AIHA/ASSE Z10–2012. This management system can be effectively implemented not only to achieve significant safety and health benefits but also to have a favorable effect on productivity, financial performance, quality, and other business goals (Manuele, 2014). For example, it has also been used in the development of a tool for analyzing the performance of the requirements of the safety management systems through the reporting and measuring of the defined Key Performance Indicators (Valdez Banda & Goerlandt, 2018). In addition to obtaining effective results in the implementation of the PDCA-based approach to Environmental-Value Stream Mapping (E-VSM), it can be an effective alternative to improve the green performance of operations as well (Garza-Reyes et al., 2018). Finally, PDCA has been used to strengthen the relationship between lean manufacturing and ergonomics (Nawawi et al., 2018). Accordingly, this is evidence that this cycle can be applied successfully to all processes, and its versatility can improve the quality management system as a whole, EMS, and HSMS.

In terms of health and safety standards, the most commonly used standards are OHSAS 18001 and ISO 45001 (Hoque & Shahinuzzaman, 2021). Any HSMS has some common elements which ensure health and safety for the employees and workers, such as policy, organizing, planning, implementing, evaluating, and taking actions for improvement (Yanar et al., 2020). It is widely implemented in organizations worldwide (Mohammadfam et al., 2017). Figure 1.2 shows how the

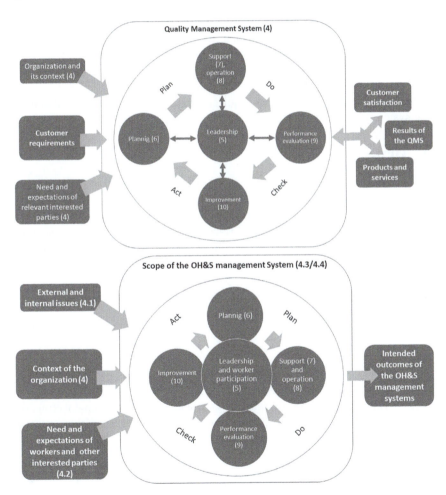

FIGURE 1.2 ISO Standard in the PDCA cycle for QMS and HSMS adapted from the quality management system is the first diagram, *ISO 9001:2015(En), Quality Management Systems— Requirements* (www.iso.org/obp/ui/#iso:std:iso:9001:ed-5:v1:en). The second diagram is the health and safety management system, *ISO 45001:2018(En), Occupational Health and Safety Management Systems—Requirements with Guidance for Use* (www.iso.org/obp/ui/#iso:std:iso:45001:ed-1:v1:en).

requirements of the ISO standards can be grouped in the PDCA cycle for both management systems.

The aforementioned literature review shows an overview of management systems, as well as existing models, and standards that can be used in the conformation of the ergonomics management model. These have been successfully implemented in several organizations in various sectors and countries (International Standard Organization, 2021). Other relevant aspects refer to their systemic approach providing an integral perspective and implementation in the organizations by determining the internal and external aspects that can influence the results, as well as their synergy and systematic process for the achievement of results. In addition, some of the advantages found are among others the full control of compliance obligations, a significant reduction in injury indexes (in the case of safety and health management), a reduction in the associated costs, and an improvement in the corporate image (Campailla et al., 2019) as well as the positive and significant effects on operational performance (Fahmi et al., 2021). Besides, one of the strongest aspects is the structure of the standards based on the ten main clauses of the ISO high-level structure. This feature leads to a high potential for integrating altogether requirements into a single integrated management system (Darabont et al., 2019).

1.3   METHODOLOGY

This research used the SLR following the PRISMA Declaration (Liberati et al., 2009), as retrieved from its website: www.prisma-statement.org/. The document describes the sources of information, the search parameters in the databases, the refinement of the results, the final selection of the identified findings, and an analysis of the results. Figure 1.3 shows the five stages in which the process was structured.

1.4   RESULTS

The results obtained during each stage of the SLR are shown in the following sections.

1.4.1   DATABASE SELECTION

The search was conducted in the ScienceDirect, ProQuest, SpringerLink, and Emerald Insight databases as these are the most widely used in the engineering, supply chain, safety, and ergonomics fields, according to the analysis of other systematic literature reviews associated with this topic.

FIGURE 1.3   Stages of the literature review process for this research.

1.4.2   IDENTIFICATION OF SEARCH PARAMETERS

The scope for the SLR for all databases covered journal articles published between 2010 and 2021, which featured keywords in both their title and content. Logical operators were also considered. Both parameters used were ("ergonomics management systems" OR "safety management systems" OR "quality management systems") AND ("supply chain").

Inclusion criteria:

1. The paper is published in a scientific journal.
2. The paper is available in English.
3. The paper reports on management systems related to the prevention of injuries and health impairment of personnel in work activities to provide better workplace designs.
4. The paper reports on safety or ergonomics or quality management systems used in the supply chain.

Exclusion criteria:

1. Duplicated papers
2. Papers include conference posters, abstracts, short papers, and unpublished works.
3. Papers to address the food safety management systems (SMS) in the supply chain.
4. Papers fail to address the management systems in the supply chain.

1.4.3   PAPER SELECTION PROCESS

A total of 1,135 articles were found after the initial search through the four databases (ScienceDirect, ProQuest, SpringerLink, and Emerald Insight). Figure 1.4 shows the selection process, as well as the results of the screening once the selection and exclusion criteria were established. The results of each stage in the selection process are also featured. It should be noted that the screening process was based on the analysis of all papers in their entirety.

1.4.4   RESULTS OF FINAL SELECTION

Figure 1.4 shows the 39 papers that met the inclusion criteria for the final selection. Next, Table 1.1 shows the total of articles selected, which were in turn classified by year of publication, author, management systems implemented or evaluated, standards, and management models.

1.4.5   ANALYSIS OF RESULTS

As a result of the process mentioned earlier, three different management systems, three standards, and three management models were identified in the 39 articles

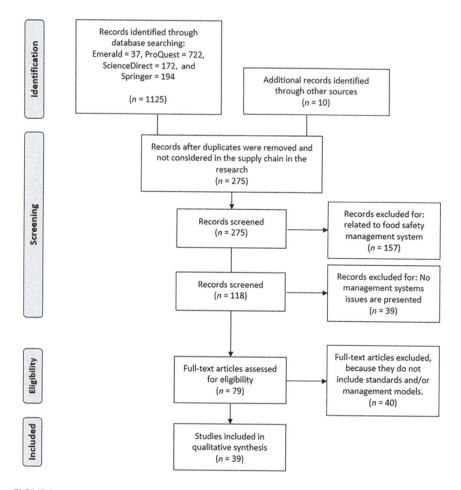

FIGURE 1.4 Paper selection process.

selected. These can be the basis for proposing an ergonomics management model. As seen in Figure 1.5, the analyzed literature showed a growing interest in the subject studied since publications addressing it increased fivefold during the established period. For the last eight years, the annual number of publications has been above two, while in previous years, the annual average was 0.5 publications per year. This suggests that the topic is far from being exhausted, and its popularity among researchers is still rising. Therefore, it is safe to say that further unique studies regarding this field of knowledge will continue to appear soon.

Another relevant aspect refers to the papers' most studied management systems. Figure 1.6 shows the 39 articles classified by percentages according to the systems type. As can be seen, an area of opportunity lies in developing investigations related to the implementation, evaluation, or design of the ergonomics management system (EMS). However, one of the most common forms of intervention related to this issue within organizations is the implementation of ergonomics programs to fit the job

TABLE 1.1

Characteristics of the Selected Articles

Year	Author/s	Management Systems			Standards and Management Models					
		HSMS or SMS	QMS	EMS	ISO 9001	ISO 45001	OSHAS 18001	EFQM	TQM	PDCA
2011	(Liu et al., 2011)	X			X					X
2011	(Bayo-Moriones et al., 2011)		X		X			X	X	
2014	(Poli et al., 2014)									X
2014	(Wu et al., 2014)	X								X
2014	(McGuinness & Utne, 2014)	X	X		X		X			
2014	(Ferreira Rebelo et al., 2014)	X	X		X		X			X
2015	(Asgher et al., 2015)		X		X			X		
2015	(Kafetzopoulos et al., 2015)		X							
2016	(dos Santos et al., 2016)			X						X
2016	(Mohammadfam et al., 2016)	X					X			X
2017	(Zimon, 2017a)		X		X					
2017	(Zimon, 2017b)								X	
2017	(Zimon, 2017c)		X		X					
2017	(Zimon & Malindžák, 2015)		X		X					
2017	(Fonseca & Domingues, 2017)		X		X					
2017	(Suárez et al., 2017)							X	X	
2017	(Sadegh Amalnick & Zarrin, 2017a)	X	X	X				X		
2018	(Hohnen & Hasle, 2018)	X					X			X
2018	(Muhamad Khair et al., 2018)	X	X		X		X			
2018	(Hallberg et al., 2018)		X		X					

(*Continued*)

TABLE 1.1　(Continued)
Characteristics of the Selected Articles

Year	Author/s	Management Systems			Standards and Management Models					
		HSMS or SMS	QMS	EMS	ISO 9001	ISO 45001	OSHAS 18001	EFQM	TQM	PDCA
2019	(Varella & Trindade, 2019)			X						X
2019	(Refaat & El-Henawy, 2019)		X		X					
2019	(Morgado et al., 2019)	X				X				
2019	(Çalış & Buyükakinci, 2019b)	X								
2019	(da Silva & Amaral, 2019)	X				X	X			
2019	(El Manzani et al., 2019)		X		X					
2020	(Zimon et al., 2020)		X		X				X	
2020	(Uhrenholdt Madsen et al., 2020)	X				X	X			
2020	(Swuste et al., 2020)	X							X	X
2021	(Grijalvo & Sanz-Samalea, 2020)		X		X					
2021	(Medina-Serrano et al., 2021)		X		X					X
2021	(García-Aranda et al., 2021)		X					X	X	
2021	(Markowski et al., 2021)	X				X	X			X
2021	(Rudakov et al., 2021)	X				X	X			X
2021	(L. Fonseca et al., 2021)							X	X	
2021	(Haas & Yorio, 2021)	X				X	X			
2021	(Tebar Betegon et al., 2021)		X		X					X
2021	(Hoque & Shahinuzzaman, 2021)	X				X	X			
2021	(Bagodi et al., 2021)		X		X					

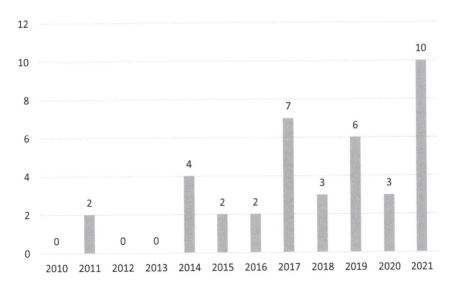

FIGURE 1.5 Year of publication of final selections.

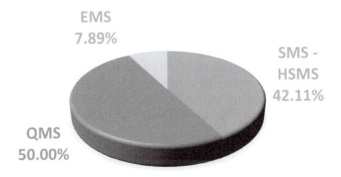

FIGURE 1.6 The proportion of the evaluated management system on the supply chain.

to the worker's abilities and capacities by reducing the physical and mental workload. Unfortunately, these programs sometimes lack a proactive and participatory approach (Fernandes et al., 2015). In addition, most of the time, these are considered separate initiatives by top management and struggle to get a priority place at the table as product quality, worker safety, and profitability have.

Furthermore, a key element found in the literature is ergonomics auditing, as it is used to track improvements and provide a standard for the ideal and mature program. Alpaugh-Bishop (2012) considers that the key elements for the success of the program should be audited as follows:

- Management commitment/foundation for success/program infrastructure
- Ergonomics training/awareness

- Identifying problematic jobs/understanding MSD hazards/ergonomics analyses
- Selecting ergonomics solutions/implementing solutions/communicating success
- Health care management/return to work/physical demands descriptions
- Proactive ergonomics/design ergonomics

Ergonomics program managers must overcome obstacles to effectively sustain results and demonstrate how ergonomics initiatives fit naturally with the organization's continuous improvement philosophies (Monroe et al., 2012). Under this context, organizations need to integrate a systematic ergonomic improvement process to identify and reduce employee exposure to risk factors. Munck-Ulfsfält et al. (2003) suggested that ergonomics is not a separate entity but a strategy which facilitates compliance.

On the other hand, the articles were analyzed considering the number of standards and management models used in each paper. Figure 1.7 shows the percentages of papers found per aspect. It can be noted that the standards most frequently used are ISO 9001 for their nature in quality management systems, while the least used is ISO 45001 in safety management systems. OSHAS 18001 will be withdrawn and organizations should migrate to ISO 45001 by March 2021, as the latter ensures greater compatibility with ISO 9001 and 14001, which facilitates the formation of an integrated management system. As for management models, PDCA is the most widely used in research related to the supply chain.

Additionally, the articles were analyzed considering the relation between standards and management models used to evaluate or design the management systems. For this purpose, their nature and adaptability are considered. The articles reviewed show a greater frequency of use of 18001 standard and the PDCA model (six articles), followed by the option of ISO 9001 and PDCA (four articles), and in the third option,

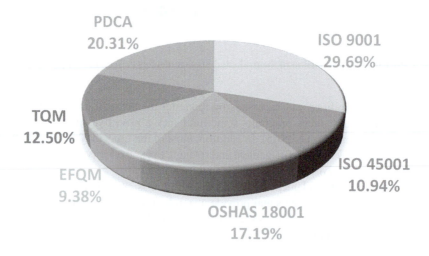

FIGURE 1.7 The proportion of the number of standards and management models used in each paper.

ISO 45001 and PDCA (three articles). The 9001 standard is related to OSHAS 18001, TQM, and EFQM, confirming that this is the standard with the greatest adaptability. The PDCA model is the most used for the development of management systems. This was confirmed by Poli et al. (2014), Haight et al. (2014), and Haas and Yorio (2016). It is important to note that the new version of ISO 9001: 2015 seeks to be less prescriptive and more focused on managing the entire supply chain, requires greater management commitment, is also less bureaucratic, and is more friendly for organizations (Zimon, 2017c). Regarding the focus of ergonomics management systems, three types of research were identified:

1. Concerning this research, dos Santos et al. (2016) studied the implementation of the Participatory ergonomics management system using TQM tools, aimed at adapting workstations, the workplace, and work conditions to comply with Brazilian Labor Legislation. This model has proven effective in identifying ergonomic problems in workstations, production processes, and working conditions and has allowed the implementation of ergonomic actions. However, there is no global application in the supply chain or an approach to macroergonomic problems, as its focus is microergonomic.

2. The purpose of this research is to present an integrated framework for the performance of the human resource (HR) with respect to the factors of health, safety, environment, and ergonomics (HSEE) management system, and also the criteria of European federation for quality management (EFQM) as one of the well-known business excellence models. A questionnaire was designed to evaluate the company's performance using the fuzzy data envelopment analysis (FDEA) and intelligent algorithm based on an adaptive neuro-fuzzy inference system (ANFIS). Furthermore, the impact of the factors on the company's performance and their strengths and weaknesses are identified by conducting a sensitivity analysis of the results. Similarly, a design of the experiment is performed to prioritize the factors. In this research, the integrated management model is not implemented. Nevertheless, it is used for the evaluation of the organization and only addresses the ergonomics aspects in three questions of the evaluation instrument.

3. Varella and Trindade (2019) presented the following as the objective of their research and management model in ergonomics that has been certified in Brazil by the representative body of ISO (International Organization for Standardization), ABNT (Brazilian Association of Technical Norms), and has been showing good results, through documented, registered, and controlled actions, respecting the PDCA cycle management system. The program involves several actions (planned, controlled, and documented) based on the ergonomics of the activity (whose main objective is to understand the work to transform it) and in the national legislation, covering the ergonomic workplace analyses (which contemplate the three dimensions of ergonomics—physical, cognitive, and organizational), execution and validation of projects of ergonomic improvements (conception and

correction), investigation of work-related out- patient complaints, follow-up of return to work processes, and inclusion of people with disabilities in workstations, training (both work-related and non-work-related issues) and actions aimed at the well-being and quality of life of the employees, with relaxation, strengthening, and postural alignment activities in a place inside the company equipped with professionals specialized in Physical Education and Physiotherapy. Therefore, with the combination of the activity's ergonomics concepts and the PDCA cycle management model, it is possible to propose steps for an ergonomics program that seeks the continuous improvement of the work processes and the constant validation of the actions performed by it. The model's main activities are as follows:

- Recording of program ergonomic procedures.
- Annual census of all existing jobs in the company, for subsequent preparation of the annual planning for the implementation of Ergonomic Work Analysis.
- Elaboration and control of visits and technical opinions requested by the medical department or by the areas of the company.
- Request, follow-up, and validation of ergonomic improvements validated by the personnel.
- Participation in the projects of new models and new workstations of the company.
- Training related to ergonomics.
- Support other programs.
- Management of the Labor Gymnastics Program and the "Ergonomics Center": space for quality of life at work, where physical activities of different modalities (Pilates classes, functional training, postural orientations, relaxation, among others) are carried out for all employees.

These articles agree that the participation of employees or specialists in ergonomics management is necessary and may be increased due to the implementation of these models. Furthermore, in the case of the Sadegh Amalnick and Zarrin (2017b) research, leadership was shown to be a significant element in system performance. Therefore, both elements should be considered in the design of the ergonomics management model. On the other hand, the lack of models or management systems for the evaluation and/or prevention of the lack of ergonomics in organizations facilitate integration with existing management models in organizations such as those related to quality, environment, health, and safety. Lewandowski (2000) emphasized the importance of integrating ergonomics as a general concept in a total quality management system. He suggested that to achieve the effects of continuous improvement of occupational health and safety and quality, ergonomics must be considered in management processes.

In the case of MSDs, Yazdani et al. (2015b) confirmed the insufficient literature describing the integration of assessment and prevention in management systems. This lack of information may isolate the prevention of MSDs, which hinders the prevention of these disorders at the organizational level. In addition, it has been observed that ergonomics activities are rarely incorporated into integrated management systems or occupational health and safety systems of companies (Caroly et al., 2010). However, De Oliveira Matias and Coelho (2002) assert that the benefits of incorporating ergonomics into different management systems could be improved by integrating ergonomics into these management systems.

HSMS is a formalized framework for organizations to manage workers' health and safety through the association of arrangements, the planning and review, and the program elements that work together to enhance safety performance. In implementing these systems in organizations, there is evidence that they can address and mitigate workplace risks (Yazdani et al., 2015a). Similarly, the purpose of ergonomics is to identify, analyze, and reduce occupational hazards by adapting the workplace and conditions to the characteristics of the operator. This affinity facilitates its integration to the HSMS and, therefore, to a standard within this system, such as the Occupational Health and Safety Assessment Series (OHSAS 18001:2007), the ANSI Z10 standard, and ISO 45001:2018, among others.

Emily J. Haas and Yorio (2021) made a comparison between the OSHAS 18001 and ANSI Z10 standards, considering as elements to compare leadership development, responsibility, and accountability; risk management; emergency management training, culture enhancement, communication, and collaboration; reinforcement and recognition; change management; resources and planning, work procedures, and permits; occupational health; incident investigation; behavior optimization; engineering and construction; contractor management; assurance, documentation, and information management, detecting the lack of leadership development, culture improvement and behavior optimization in both standards, while the particular case of ANSI Z10 lacks change management and resources and planning as aspects within its implementation. On the other hand, the authors Rostykus et al. (2016) propose integrating ergonomic aspects to ISO 45001, as this model can be used as an effective system to manage ergonomics.

Another way to compare standards and management models is through their structure or requirements that make them up; for this purpose in Table 1.2, the comparison is shown and those aspects that are lacking in terms of ISO 45001 are identified.

Regarding the comparison of ISO 45001 with ANSI Z10, we can see in the table that ISO 45001 has more requirements. At the same time, ANSI Z10 lacks an exclusive element for determining legal requirements, action planning, conformity, and assessment and does not consider an audit program. On the other hand, when comparing ISO 45001 with EFQM, we find some difficulty due to the different methodological approaches (e.g., attributes to measure, variables, measurement scales); however, we can rescue aspects related to the commitment with stakeholders such as customers, people, companies and interest groups, governors, society, partners, and suppliers, which can facilitate the inclusion of the ergonomics management system that we are seeking to develop. Even the EFQM model is considered vital to manage an organization that wants a long-term sustainable future; some approaches suggest the implementation of ISO 9001 and then using the EFQM to achieve excellence through sustainable and outstanding results (Fonseca et al., 2021).

TABLE 1.2

Comparison between QMS and HSMS

PDCA Cycle	Health and Occupational Safety Management Systems			Quality Management Systems
	ISO 45001:2018 Occupational Health and Safety Management Systems—Requirements with Guidance for Use	ANSI ASSE Z10–2012 (R2017) Occupational Health and Safety Management Systems		European Foundation for Quality Management (EFQM)
	5. Leadership and worker participation	**3.0 Management leadership and employee participation**	Direction	**1. Purpose, vision, and strategy**
	5.1 Leadership and commitment	3.1 Management Leadership—3.1.1 Occupational health and safety management system		1.1 Define purpose and vision
				1.2 Identify and understand stakeholders needs
	5.2 OH&S policy	3.1.2 OHS policy		1.3 Understand the ecosystem, own capabilities, and major challenges
				1.4 Develop strategy
	5.3 Organizational roles, responsibilities, and authorities	3.1.3 Responsibility and authority	RADAR	1.5 Design and implement a governance and performance management system
	5.4 Consultation and participation of workers	3.2 Employee participation	Approaches	**2. Organizational culture and leadership**
Plan	**6 Planning**	**4.0 Planning**		2.1 Steer the organization's culture and nurture values
	6.1 Actions to address risks and opportunities	4.1 Review process		2.2 Create the conditions for realizing change
	6.1.1 General	4.2 Assessment and prioritization		2.3 Enable creativity and innovation
	6.1.2 Hazard identification and assessment of risks and opportunities	5.1.1 Risk assessment		
	6.1.3 Determination of legal requirements and other requirements			

Do			Deploy		Execution
6.1.4 Planning action					2.4 Unite behind and engage in purpose, vision, and strategy
6.2 OH&S objectives and planning to achieve them	4.3 Objectives				**3. Engaging stakeholders**
6.2.1 OH&S objectives					3.1 Customers: build sustainable relationships
6.2.2 Planning to achieve OH&S objectives	4.4 Implementation plans				3.2 People: attract, engage, develop, and retain
					3.3 Business and governing stakeholders—secure and sustain ongoing support
					3.4 Society: contribute to development, well-being, and prosperity
					3.5 Partners and suppliers: build relationships and ensure support for creating sustainable value
7 Support			4.4 Allocation of resources		5.5 Manage assets and resources
7.1 Resources			5.2 Education, training, awareness, and competence		3.2 People: attract, engage, develop, and retain
7.2 Competence					
7.3 Awareness			5.3 Communication		4.2 Communicate and sell the value
7.4 Communication					
7.4.1 General					
7.4.2 Internal communication					

(Continued)

TABLE 1.2 (Continued)
Comparison between QMS and HSMS

PDCA Cycle	Health and Occupational Safety Management Systems		RADAR	Quality Management Systems
	ISO 45001:2018 Occupational Health and Safety Management Systems—Requirements with Guidance for Use	ANSI ASSE Z10–2012 (R2017) Occupational Health and Safety Management Systems		European Foundation for Quality Management (EFQM)
	7.4.3 External communication			
	7.5 Documented information	5.4 Document and record control process		
	7.5.1 General			
	7.5.2 Creating and updating			
	7.5.3 Control of documented information			
	8 Operation	**5.0 Implementation and operation**		**4. Creating sustainable value**
	8.1 Operational planning and control	5.1 OHSMS operational elements		4.1 Design the value and how it is created
	8.1.1 General			
	8.1.2 Eliminating hazards and reducing OH&S risks	5.1.2 Hierarchy of controls		4.3 Deliver the value
	8.1.3 Management of change	5.1.3 Design Review and management of change		4.4 Define and implement the overall experience
	8.1.4 Procurement	5.1.4 Procurement and 5.1.5 Contractors		
		5.1.6 Emergency preparedness		
	8.2 Emergency preparedness and response			
Check	**9 Performance evaluation**	**6.0 Evaluation and corrective action**	Assess and refine	**5. Driving performance and transformation**
	9.1 Monitoring, measurement, analysis, and evaluation	6.1 Monitoring, measurement, and assessment		5.1 Drive performance and manage risk
	9.1.1 General			5.2 Transform the organization for the future

			Result	Result
	9.1.2 Evaluation of compliance			5.3 Drive innovation and utilize technology
	9.2 Internal Audit	6.3 Audits		5.4 Leverage data, information, and knowledge
	9.2.1 General			**6. Stakeholder perceptions**
	9.2.2 Internal audit program			**7. Strategic and operational performance**
	9.3 Management review	**7.0 Management review**		
		7.1 Management review process		
		7.2 Management review outcomes and follow-up		
Act	**10 Improvement**	6.2 Incident investigation		
	10.1 General			
	10.2 Incident, nonconformity, and corrective action	6.4 Corrective and preventive actions		
	10.3 Continual improvement	6.5 Feedback to the planning process		

TABLE 1.3

Ergonomics Management System Constructs and Domains

PDCA Cycle	ISO 45001:2018 Occupational Health and Safety Management Systems—Requirements with Guidance for Use	
	Constructs	Domains
	5 Leadership and worker participation	5.1 Leadership and commitment
		5.2 Policy
		5.3 Organization roles, responsibilities, and authorities
		5.4 Consultation and participation of workers
Plan	6 Planning	6.1 Actions to address risks and opportunities
		6.2 Objectives and planning to achieve them
DO	7 Support	7.1 Resources
		7.2 Competence
		7.3 Awareness
		7.4 Communication
		7.5 Documented information
	8 Operation	8.1 Operational planning and control
		8.2 Emergency preparedness and response
Check	9 Performance evaluation	9.1 Monitoring, measurement, analysis, and evaluation
		9.2 Internal audit
		9.3 Management review
Act	10 Improvement	10.1 General
		10.2 Incident, nonconformity, and corrective action
		10.3 Continual improvement

Considering these findings, the decision was made to adopt ISO 45001 as a structural element and the PDCA cycle, both of which are relevant and considered key to the development of the EMS. As for the domains to be considered in the ergonomics management system, these are shown in Table 1.3; these domains are classified for each construct and under the PDCA cycle. In addition, leadership and worker participation are integrated into the model.

1.5 THE OPPORTUNITY OF ISO 45001 FOR ERGONOMICS MANAGEMENT IN THE SUPPLY CHAIN

From the initiative stage, ISO 45001 sought to ensure a "robust and effective set of processes to improve occupational safety in global supply chains" (Hemphill & Kelly, 2016); in other words, a sustainable solution to promote occupational health and safety in global supply chains. This standard is based on the PDCA model; iterative process organizations use to achieve continuous improvement. Therefore, it can be explored as a value-added capability for global supply chains. The proposed structure in ISO 45001 associated with safety management systems provides a common framework and terminology for managing hazards in the workplace. This same framework can be applied to identify systematically, control, and verify the reduction of related risk factors.

Rostykus et al. (2016) state that aligning how the organization addresses ergonomics through a management system allows Occupational Safety and Health professionals to communicate and engage business leaders in a way they are already familiar with. In addition, they confirm that ISO 45001 is a model that can be used as an effective system for ergonomics management. Additionally, the opportunities of ISO 45001 to contribute to proposing an ergonomics management model are identifying hazards, communicating them, and addressing the analysis and mitigation of known hazards. In addition, there are other opportunities related to system improvement; within these are the identified ergonomic assessments and other injury prevention assessments (ISO 45001, 2018). Thus, to perform a practical implementation, it is important to consider whether starting an EMS from scratch or developing from an existing program, they are essential for success:

1. Evaluate the current ergonomics program/process based on a management system model.
2. Define common goals, measures, requirements, roles and responsibilities, and standard tools in a baseline document on which all department and site ergonomic improvement processes are based.
3. Obtain the buy-in, sponsorship, and participation from key leaders.
4. Implement the ergonomic improvement process at each location or department through the sponsor, subject matter experts, and engineers. Ensure they use standard assessment tools for consistent reporting and tracking, and share practical improvements and best practices. Track progress and metrics regularly.
5. Audit each site/department's ergonomics management system to ensure compliance with business requirements; identify best practices and opportunities for improvement; and engage leadership to refine their plans and focus on sustaining the process.

1.6 CONCLUSIONS

This chapter shows an overview of the management systems used in the SC, through a literature review covering the period from 2010 to 2021. The methodology followed the PRISMA Guidelines and proved effective in identifying the models and standards of the quality, health and safety, and ergonomics management systems, as well as their domains and features. As confirmed, the objective of this research was met to establish the basis for designing an ergonomics management model that allows the evaluation of companies.

Thirty-nine researches were identified that met the inclusion criteria, of which 50% were related to QMS. These systems are the most used for evaluation and implementation in organizations. From the SLR, leadership, employee participation, and the audit of compliance with the management system requirements are considered fundamental elements for the design of the EMS.

In addition, the ISO 9001 standard and the PDCA cycle stand out; however, ISO 45001 was chosen as the basis for the EMS since it is compatible with ISO 9001, it is also based on the PDCA cycle, and due to its nature, ergonomics is highly related to health and safety management systems, since both focus on risk analysis from its field of action. Not forgetting that within its regulatory framework, it establishes an opportunity for inclusion and thus improves the working conditions and health of the worker.

The purpose of an ergonomics management system is to contribute to the social sustainability of the SC through an evaluation system that provides an ergonomics management index, which gives an overview of the management level of the practices adopted in each element of the SC, as well as globally throughout the SC. This will require the integration of other key elements such as corporate social responsibility and collaboration, cooperation, and coordination between the SC links that comprise it since through these practices, the interaction of the links is enabled, and better performance of the sustainable SC is obtained (Dias & Silva, 2022).

REFERENCES

Alpaugh-Bishop, A. L. (2012). *Building and using an ergonomics audit: Does your program make the grade?* www.taylordergo.com/wp-content/uploads/2012/10/building-and-using-an-ergonomics-audit-ACE-2009.pdf

Asgher, U., Leba, M., Ionică, A., Moraru, R. I., & Ahmad, R. (2015). Human factors in the context of excellence models: European foundation for quality management (EFQM) excellence software model and cross-cultural analysis. *Procedia Manufacturing, 3*, 1758–1764. https://doi.org/10.1016/J.PROMFG.2015.07.479

Bagodi, V., Thimmappa Venkatesh, S., & Sinha, D. (2021). A study of performance measures and quality management in small and medium enterprises in India. *Benchmarking: An International Journal, 28*(4), 1356–1389. https://doi.org/10.1108/BIJ-08-2020-0444

Bayo-Moriones, A., Merino-Díaz-De-Cerio, J., Antonio Escamilla-De-León, S., & Mary Selvam, R. (2011). The impact of ISO 9000 and EFQM on the use of flexible work practices. *International Journal of Production Economics, 130*(1), 33–42. https://doi.org/10.1016/J.IJPE.2010.10.012

Bhasi, M., Hisham, H., & Vinodkumar, M. N. (2010). Safety management practices and safety behaviour: Assessing the mediating role of safety knowledge and motivation related papers safety management practices and safety behaviour: Assessing the mediating role of safety knowledge and motivation. *Accident Analysis and Prevention, 42*, 2082–2093. https://doi.org/10.1016/j.aap.2010.06.021

Çalış, S., & Buyükakinci, B. Y. (2019a). Occupational health and safety management systems applications and a system planning model. *Procedia Computer Science, 158*, 1058–1066. https://doi.org/10.1016/j.procs.2019.09.147

Çalış, S., & Buyükakinci, B. Y. (2019b). Occupational health and safety management systems applications and a system planning model. *Procedia Computer Science, 158*, 1058–1066. https://doi.org/10.1016/J.PROCS.2019.09.147

Campailla, C., Martini, A., Minini, F., & Sartor, M. (2019). ISO 45001. In *Quality management: Tools, methods and standards* (pp. 217–243). Emerald Group Publishing Ltd. https://doi.org/10.1108/978-1-78769-801-720191014

Caroly, S., Coutarel, F., Landry, A., & Mary-Cheray, I. (2010). Sustainable MSD prevention: Management for continuous improvement between prevention and production. Ergonomic intervention in two assembly line companies. *Applied Ergonomics, 41*(4), 591–599. https://doi.org/10.1016/J.APERGO.2009.12.016

Ciccarelli, M., Papetti, A., Cappelletti, F., Brunzini, A., & Germani, M. (2022). Combining world class manufacturing system and industry 4.0 technologies to design ergonomic manufacturing equipment. *International Journal on Interactive Design and Manufacturing*, *16*(1), 263–279. https://doi.org/10.1007/S12008-021-00832-7/FIGURES/9

Darabont, D. C., Bejinariu, C., Baciu, C., & Bernevig-Sava, M. A. (2019). Modern approaches in integrated management systems of quality, environmental and occupational health and safety. *Quality—Access to Success*, *20*, 105–108.

da Silva, S. L. C., & Amaral, F. G. (2019). Critical factors of success and barriers to the implementation of occupational health and safety management systems: A systematic review of literature. *Safety Science*, *117*, 123–132. https://doi.org/10.1016/j.ssci.2019.03.026

Deighton, M. G., & Deighton, M. G. (2016). Chapter 5—maintenance management. *Facility Integrity Management*, 87–139.

De Oliveira Matias, J. C., & Coelho, D. A. (2002). The integration of the standards systems of quality management, environmental management and occupational health and safety management. *International Journal of Production Research*, *40*(15 spec.), 3857–3866. https://doi.org/10.1080/00207540210155828

Dias, G. P., & Silva, M. E. (2022). Revealing performance factors for supply chain sustainability: A systematic literature review from a social capital perspective. *Brazilian Journal of Operations & Production Management*, *19*(1), 1–18.

Disterheft, A., Ferreira Da Silva Caeiro, S. S., Ramos, M. R., & De Miranda Azeiteiro, U. M. (2012). Environmental management systems (EMS) implementation processes and practices in European higher education institutions—top-down versus participatory approaches. *Journal of Cleaner Production*, *31*, 80–90. https://doi.org/10.1016/j.jclepro.2012.02.034

Domingues, P., Sampaio, P., & Arezes, P. M. (2016). Integrated management systems assessment: A maturity model proposal. *Journal of Cleaner Production*, *124*, 164–174. https://doi.org/10.1016/jjclepro.2016.02.103

dos Santos, C. M. D., Santos, R. F., Santos, A. F., & de Castro Moreira Rosa, M. (2016). Participatory ergonomics management in a textile thread plant in Brazil employing total quality management (TQM) tools. *Advances in Intelligent Systems and Computing*, *485*, 277–288. https://doi.org/10.1007/978-3-319-41983-1_25

El Manzani, Y., Sidmou, M. L., & Cegarra, J. (2019). Does ISO 9001 quality management system support product innovation? An analysis from the sociotechnical systems theory. *International Journal of Quality & Reliability Management*, *36*(6), 951–982. https://doi.org/10.1108/IJQRM-09-2017-0174

Fahmi, K., Mustofa, K., Rochmad, I., Sulastri, E., Sri Wahyuni, I., & Irwansyah, I. S. (2021). Effect ISO 9001:2015, ISO 14001:2015 and ISO 45001:2018 on operational performance of automotive industries. *Journal of Industrial Engineering & Management Research*, *2*(1996), 6.

Fernandes, P. R., Hurtado, A. L. B., & Batiz, E. C. (2015). Ergonomics management with a proactive focus. *Procedia Manufacturing*, *3*, 4509–4516. https://doi.org/10.1016/j.promfg.2015.07.465

Ferreira Rebelo, M., Santos, G., & Silva, R. (2014). A generic model for integration of quality, environment and safety management systems. *The TQM Journal*, *26*(2), 143–159. https://doi.org/10.1108/TQM-08-2012-0055

Fonseca, L. M., Amaral, A., & Oliveira, J. (2021). Quality 4.0: The EFQM 2020 model and industry 4.0 relationships and Implications. *Sustainability*, *13*(6), 3107. https://doi.org/10.3390/SU13063107

Fonseca, L. M., & Domingues, J. P. (2017). Reliable and flexible quality management systems in the automotive industry: Monitor the context and change effectively. *Procedia Manufacturing*, *11*(June), 1200–1206. https://doi.org/10.1016/j.promfg.2017.07.245

García-Aranda, J. R., Ortega-Lapiedra, R., & Bernués-Olivan, J. (2021). Sustainability, efficiency, and competitiveness in rail mobility: The ADIF-Spain case study. *Sustainability*, *13*(16), 8977. https://doi.org/10.3390/SU13168977

Garza-Reyes, J. A., Torres Romero, J., Govindan, K., Cherrafi, A., & Ramanathan, U. (2018). A PDCA-based approach to environmental value stream mapping (E-VSM). *Journal of Cleaner Production*, *180*, 335–348. https://doi.org/10.1016/j.jclepro.2018.01.121

Goethe, D., Romero, J., & Romero, J. (2022). Industry 4.0 for sustainable supply chain management: Drivers and barriers. *Procedia Computer Science*. https://doi.org/10.1016/j.procs.2022.07.094

Gómez-López, R., Serrano-Bedia, A. M., & López-Fernández, M. C. (2019). An exploratory study of the results of the implementation of EFQM in private Spanish firms. *International Journal of Quality and Reliability Management*, *36*(3), 331–346. https://doi.org/10.1108/IJQRM-01-2018-0023

Grijalvo, M., & Sanz-Samalea, B. (2020). Exploring EN 9100: Current key results and future opportunities—a study in the Spanish aerospace industry. *Economic Research-Ekonomska Istraživanja*, *34*(1), 2712–2728. https://doi.org/10.1080/13316 77X.2020.1838312; http://www.tandfonline.com/action/authorsubmission?journalcode= rero20&page=instructions

Gunawan, M., Asyahira, R., & M Sidjabat, F. (2020). Environmental management system implementation in MSMEs: A literature review. *Jurnal Serambi Engineering*, *5*(2), 1070–1078. https://doi.org/10.32672/jse.v5i2.1958

Haas, E. J., & Yorio, P. L. (2016). Exploring the state of health and safety management system performance measurement in mining organizations. *Safety Science*, *83*, 48–58. https://doi.org/10.1016/J.SSCI.2015.11.009

Haas, E. J., & Yorio, P. L. (2021). Exploring the differences in safety climate among mining sectors. *Mining, Metallurgy and Exploration*, *38*(1), 655–668. https://doi.org/10.1007/s42461-020-00364-w

Haight, J. M., Yorio, P., Rost, K. A., & Willmer, D. R. (2014). Safety management systems—comparing content & impact. *Professional Safety*, *59*(5), 44–51.

Hallberg, P., Hasche, N., Kask, J., & Öberg, C. (2018). Quality management systems as indicators for stability and change in customer-supplier relationships. *IMP Journal*, *12*(3), 483–497. https://doi.org/10.1108/IMP-01-2018-0006

Hemphill, T. A., & Kelly, K. J. (2016). Socially responsible global supply chains: The human rights promise of shared responsibility and ISO 45001. *Journal of Global Responsibility*, *34*(1), 1–5.

Hohnen, P., & Hasle, P. (2018). Third party audits of the psychosocial work environment in occupational health and safety management systems. *Safety Science*, *109*, 76–85. https://doi.org/10.1016/j.ssci.2018.04.028

Hong, J., Zhang, Y., & Ding, M. (2018). Sustainable supply chain management practices, supply chain dynamic capabilities, and enterprise performance. *Journal of Cleaner Production*, *172*, 3508–3519. https://doi.org/10.1016/j.jclepro.2017.06.093

Hoque, I., & Shahinuzzaman, M. (2021). Task performance and occupational health and safety management systems in the garment industry of Bangladesh. *International Journal of Workplace Health Management*, *14*(4), 369–385. https://doi.org/10.1108/IJWHM-09-2020-0169

Ifadiana, D. P., & Soemirat, J. (2016). An analysis of the effect of the implementation of an integrated management system (IMS) on work ergonomics in an O&M power plant company. *Journal of Engineering and Technological Sciences*, *48*(2), 173–182. https://doi.org/10.5614/j.eng.technol.sci.2016.48.2.4

Ingason, H. T. (2015). Best project management practices in the implementation of an ISO 9001 quality management system. *Procedia—Social and Behavioral Sciences*, *194*(October 2014), 192–200. https://doi.org/10.1016/j.sbspro.2015.06.133

International Standard Organization. (2021). *The ISO survey of management system standard certifications-2020-explanatory note background*. https://qi4d.org/2020/09/16/qi-data-the-iso-survey-of-management-system-standard-certifications/

ISO. (2018). *ISO 45001:2018(en), occupational health and safety management systems—requirements with guidance for use*. www.iso.org/obp/ui/#iso:std:iso:45001:ed-1:v1:en

ISO. (n.d.). *ISO 9001:2015(en), quality management systems—requirements*. Retrieved August 1, 2022, from www.iso.org/obp/ui/#iso:std:iso:9001:ed-5:v1:en

ISO 45001. (2018). ISO 45001 Sistemas de administración/gestión en seguridad y salud ocupacional—requerimientos con guías para uso. *Secretaría Central de ISO En Ginebra, Suiza*, *1*, 1–60.

ISO 9001. (2008). *ISO 9001:2008(en), quality management systems—requirements*. www.iso.org/obp/ui/#iso:std:iso:9001:ed-4:v2:en

Kafetzopoulos, D. P., Psomas, E. L., & Gotzamani, K. D. (2015). The impact of quality management systems on the performance of manufacturing firms. *International Journal of Quality & Reliability Management*, *32*(4), 381–399. https://doi.org/10.1108/IJQRM-11-2013-0186

Korkulu, S., & Bóna, K. (2019). Ergonomics as a social component of sustainable lot-sizing: A review. In *Periodica polytechnica social and management sciences* (Vol. 27, Issue 1, pp. 1–8). Budapest University of Technology and Economics. https://doi.org/10.3311/PPso.12286

Labodová, A. (2004). Implementing integrated management systems using a risk analysis based approach. *Journal of Cleaner Production*, *12*(6), 571–580. https://doi.org/10.1016/j.jclepro.2003.08.008

Lewandowski, J. (2000). Ergonomics in total quality management. *Proceedings of the Human Factors and Ergonomics Society Annual Meeting*, *44*(July 1), 284–287. https://doi.org/10.1177/154193120004401041

Liberati, A., Altman, D. G., Tetzlaff, J., Mulrow, C., Gøtzsche, P. C., Ioannidis, J. P. A., Clarke, M., Devereaux, P. J., Kleijnen, J., & Moher, D. (2009). The PRISMA statement for reporting systematic reviews and meta-analyses of studies that evaluate health care interventions: Explanation and elaboration. *PLoS Medicine*, *6*(7). https://doi.org/10.1371/journal.pmed.1000100

Lima Marcos, E. De, Borges Silva, M., & Estevam De Souza, J. P. (2018). The integrated management system (IMS) and ergonomics: An exploratory research of qualitative perception in the application of NR-17. *Journal of Ergonomics*, *8*(3), 8–10. https://doi.org/10.4172/2165-7556.1000231

Liu, Y. J., Cao, Q. G., Wang, W. C., Tian, Z. C., & Huang, D. M. (2011). Application and development of modern safety management system in metallic and non-metallic mine. *Procedia Engineering*, *26*, 1658–1666. https://doi.org/10.1016/J.PROENG.2011.11.2351

Maldonado-Macías, A. A., Alférez-Padrón, C. R., Barajas-bustillos, M. A., Armenta-Hernández, O. D., Vargas, A. R., & Balderrama-Armendáriz, C. O. (2021). Ergonomics implementation in manufacturing industries: Management commitment for financial benefits. In *New perspectives on applied industrial ergonomics* (June, pp. 125–156). Springer International Publishing. https://doi.org/10.1007/978-3-030-73468-8

Manuele, F. A. (2014). ANSI/AIHA/ASSE Z10–2012 an overview of the occupational health & safety management systems standard. *Professional Safety* (April). www.asse.org

Markowski, A. S., Krasławski, A., Vairo, T., & Fabiano, B. (2021). Process safety management quality in industrial corporation for sustainable development. *Sustainability*, *13*(16), 9001. https://doi.org/10.3390/SU13169001

McGuinness, E., & Utne, I. B. (2014). A system engineering approach to implementation of safety management systems in the Norwegian fishing fleet. *Reliability Engineering and System Safety*, *121*, 221–239. https://doi.org/10.1016/J.RESS.2013.08.002

Medina-Serrano, R., González-Ramírez, R., Gasco-Gasco, J., & Llopis-Taverner, J. (2021). How to evaluate supply chain risks, including sustainable aspects? A case study from the German industry. *Journal of Industrial Engineering and Management, 14*(2), 120–134. https://doi.org/10.3926/jiem.3175

Mohammadfam, I., Kamalinia, M., Momeni, M., Golmohammadi, R., Hamidi, Y., & Soltanian, A. (2016). Developing an integrated decision making approach to assess and promote the effectiveness of occupational health and safety management systems. *Journal of Cleaner Production, 127*, 119–133. https://doi.org/10.1016/J.JCLEPRO.2016.03.123

Mohammadfam, I., Kamalinia, M., Momeni, M., Golmohammadi, R., Hamidi, Y., & Soltanian, A. (2017). Evaluation of the quality of occupational health and safety management systems based on key performance indicators in certified organizations. *Safety and Health at Work, 8*(2), 156–161. https://doi.org/10.1016/J.SHAW.2016.09.001

Monroe, K., Fick, F., & Joshi, M. (2012). Successful integration of ergonomics into continuous improvement initiatives. *Work, 41*(suppl.1), 1622–1624. https://doi.org/10.3233/WOR-2012-0362-1622

Morgado, L., Silva, F. J. G., & Fonseca, L. M. (2019). Mapping occupational health and safety management systems in Portugal: Outlook for ISO 45001:2018 adoption. *Procedia Manufacturing, 38*, 755–764. https://doi.org/10.1016/J.PROMFG.2020.01.103

Muhamad Khair, N. K., Lee, K. E., Mokhtar, M., & Goh, C. T. (2018). Integrating responsible care into quality, environmental, health and safety management system: A strategy for Malaysian chemical industries. *Journal of Chemical Health and Safety, 25*(5), 10–18. https://doi.org/10.1016/J.JCHAS.2018.02.003

Munck-Ulfsfält, U., Falck, A., Forsberg, A., Dahlin, C., & Eriksson, A. (2003). Corporate ergonomics programme at Volvo Car Corporation. *Applied Ergonomics, 34*(1), 17–22. https://doi.org/10.1016/S0003-6870(02)00079-0

Naeini, H. S., Dalal, K., Mosaddad, S. H., & Karuppiah, K. (2018). Economic effectiveness of ergonomics interventions. *International Journal of Industrial Engineering and Production Research, 29*(3), 261–276. https://doi.org/10.22068/ijiepr.29.3.261

Nawawi, A., Amin, M., Hasrulnizzam, W., Mahmood, W., Kamat, S. R., & Abdullah, I. (2018). Conceptual framework of lean ergonomics for assembly process: PDCA approach. *Journal of Engineering and Science Research, 2*(1), 51–62. https://doi.org/10.26666/rmp.jesr.2018.1.9

Nunhes, T. V., Bernardo, M., & Oliveira, O. J. (2019). Guiding principles of integrated management systems: Towards unifying a starting point for researchers and practitioners. *Journal of Cleaner Production, 210*, 977–993. https://doi.org/10.1016/j.jclepro.2018.11.066

Pacana, A., & Ulewicz, R. (2017). Research of determinants motiving to implement the environmental management system. *Polish Journal of Management Studies, 16*(1), 165–174. https://doi.org/10.17512/pjms.2017.16.1.14

Pereira Da Silva, M., Pruffer, C., & Amaral, F. G. (2012). Is there enough information to calculate the financial benefits of ergonomics projects? *Work, 41*(suppl.1), 476–483. https://doi.org/10.3233/WOR-2012-0199-476

Poli, M., Petroni, D., Berton, A., Campani, E., Felicini, C., Pardini, S., Menichetti, L., & Salvadori, P. A. (2014). The role of quality management system in the monitoring and continuous improvement of GMP-regulated short-lived radiopharmaceutical manufacture. *Accreditation and Quality Assurance, 19*(5), 343–354. https://doi.org/10.1007/s00769-014-1070-7

Psomas, E., & Kafetzopoulos, D. (2014). Performance measures of ISO 9001 certified and non-certified manufacturing companies. *Benchmarking, 21*(5), 756–774. https://doi.org/10.1108/BIJ-04-2012-0028

Refaat, R., & El-Henawy, I. M. (2019). Innovative method to evaluate quality management system audit results' using single value neutrosophic number. *Cognitive Systems Research, 57*, 197–206. https://doi.org/10.1016/j.cogsys.2018.10.014

Rodríguez-Mantilla, J. M., Martínez-Zarzuelo, A., & Fernández-Cruz, F. J. (2020). Do ISO:9001 standards and EFQM model differ in their impact on the external relations and communication system at schools? *Evaluation and Program Planning, 80*. https://doi.org/10.1016/j.evalprogplan.2020.101816

Rostykus, W., Ip, W., & Dustin, J. A. (2016). Managing ergonomics: Applying ISO 45001 as a model. *Professional Safety, 61*(12), 34–42. www.asse.org

Rudakov, M., Gridina, E., & Kretschmann, J. (2021). Risk-based thinking as a basis for efficient occupational safety management in the mining industry. *Sustainability, 13*(2), 470. https://doi.org/10.3390/SU13020470

Sadegh Amalnick, M., & Zarrin, M. (2017a). Performance assessment of human resource by integration of HSE and ergonomics and EFQM management system: A fuzzy-based approach. *International Journal of Health Care Quality Assurance, 30*(2), 160–174. https://doi.org/10.1108/IJHCQA-06-2016-0089

Sadegh Amalnick, M., & Zarrin, M. (2017b). Performance assessment of human resource by integration of HSE and ergonomics and EFQM management system: A fuzzy-based approach. *International Journal of Health Care Quality Assurance, 30*(2), 160–174. https://doi.org/10.1108/IJHCQA-06-2016-0089

Santos, G., Barros, S., Mendes, F., & Lopes, N. (2013). The main benefits associated with health and safety management systems certification in Portuguese small and medium enterprises post quality management system certification. *Safety Science, 51*(1), 29–36. https://doi.org/10.1016/j.ssci.2012.06.014

Seuring, S. (2012). A review of modeling approaches for sustainable supply chain management. *Decision Support Systems, 54*(4), 1513–1520. https://doi.org/10.1016/j.dss.2012.05.053

Sfreddo, L. S., Vieira, G. B. B., Vidor, G., & Santos, C. H. S. (2021). ISO 9001 based quality management systems and organisational performance: A systematic literature review. *Total Quality Management and Business Excellence, 32*(3–4), 389–409. https://doi.org/10.1080/14783363.2018.1549939

Simões, M., Carvalho, A., de Freitas, C. L., & Barbósa-Póvoa, A. (2014). How to assess social aspects in supply chains? In *Computer aided chemical engineering* (Vol. 34, pp. 801–806). Elsevier B.V. https://doi.org/10.1016/B978-0-444-63433-7.50118-8

Suárez, E., Calvo-Mora, A., Roldán, J. L., & Periáñez-Cristóbal, R. (2017). Quantitative research on the EFQM excellence model: A systematic literature review (1991–2015). *European Research on Management and Business Economics, 23*(3), 147–156. https://doi.org/10.1016/J.IEDEEN.2017.05.002

Sultan-Taïeb, H., Parent-Lamarche, A., Gaillard, A., Stock, S., Nicolakakis, N., Hong, Q. N., Vezina, M., Coulibaly, Y., Vézina, N., & Berthelette, D. (2017). Economic evaluations of ergonomic interventions preventing work-related musculoskeletal disorders: A systematic review of organizational-level interventions. *BMC Public Health, 17*(1), 935. https://doi.org/10.1186/s12889-017-4935-y

Swuste, P., van Gulijk, C., Groeneweg, J., Guldenmund, F., Zwaard, W., & Lemkowitz, S. (2020). Occupational safety and safety management between 1988 and 2010: Review of safety literature in English and Dutch language scientific literature. In *Safety science* (Vol. 121, pp. 303–318). Elsevier B.V. https://doi.org/10.1016/j.ssci.2019.08.032

Tebar Betegon, M. A., Baladrón González, V., Bejarano Ramírez, N., Martínez Arce, A., Rodríguez De Guzmán, J., & Redondo Calvo, F. J. (2021). Quality management system implementation based on lean principles and ISO 9001:2015 standard in an advanced simulation centre. *Clinical Simulation in Nursing, 51*, 28–37. https://doi.org/10.1016/j.ecns.2020.11.002

Tompa, E., De Oliveira, C., Dolinschi, R., & Irvin, E. (2008). A systematic review of disability management interventions with economic evaluations. *Journal of Occupational Rehabilitation, 18*(1), 16–26. https://doi.org/10.1007/s10926-007-9116-x

Uhrenholdt Madsen, C., Kirkegaard, M. L., Dyreborg, J., & Hasle, P. (2020). Making occupational health and safety management systems 'work': A realist review of the OHSAS 18001 standard. *Safety Science, 129.* https://doi.org/10.1016/J.SSCI.2020.104843

Valdez Banda, O. A., & Goerlandt, F. (2018). A STAMP-based approach for designing maritime safety management systems. *Safety Science, 109,* 109–129. https://doi.org/10.1016/J.SSCI.2018.05.003

Varella, C. M. C., & Trindade, M. A. L. (2019). Ergonomics management program: Model and results. *Advances in Intelligent Systems and Computing, 825,* 240–246. https://doi.org/10.1007/978-3-319-96068-5_27

Wu, B., Xu, Z., Zhou, Y., Peng, Y., & Yu, Z. (2014). Study on coal mine safety management system based on 'hazard, latent danger and emergency responses'. *Procedia Engineering, 84,* 172–177. https://doi.org/10.1016/J.PROENG.2014.10.423

Yanar, B., Robson, L. S., Tonima, S. K., & Amick, B. C. (2020). Understanding the organizational performance metric, an occupational health and safety management tool, through workplace case studies. *International Journal of Workplace Health Management, 13*(2), 117–138. https://doi.org/10.1108/IJWHM-09-2018-0126

Yazdani, A., Neumann, W. P., Imbeau, D., Bigelow, P., Pagell, M., Theberge, N., Hilbrecht, M., & Wells, R. (2015a). How compatible are participatory ergonomics programs with occupational health and safety management systems? *Scandinavian Journal of Work, Environment and Health, 41*(2), 111–123. https://doi.org/10.5271/sjweh.3467

Yazdani, A., Neumann, W. P., Imbeau, D., Bigelow, P., Pagell, M., & Wells, R. (2015b). Prevention of musculoskeletal disorders within management systems: A scoping review of practices, approaches, and techniques. *Applied Ergonomics, 51,* 255–262. https://doi.org/10.1016/j.apergo.2015.05.006

Zimon, D. (2017a). The impact of quality management systems on the effectiveness of food supply chains. *TEM Journal, 6*(4), 693–698. https://doi.org/10.18421/TEM64-07

Zimon, D. (2017b). The impact of TQM philosophy for the improvement of logistics processes in the supply chain. *International Journal for Quality Research, 11*(1), 3–16. https://doi.org/10.18421/IJQR11.01-01

Zimon, D. (2017c). The influence of quality management systems for improvement of logistics supply in Poland. *Oeconomia Copernicana, 8*(4), 643–655. https://doi.org/10.24136/oc.v8i4.39

Zimon, D., Madzik, P., & Sroufe, R. (2020). The influence of ISO 9001 & ISO 14001 on sustainable supply chain management in the textile industry. *Sustainability (Switzerland), 12*(10), 1–19. https://doi.org/10.3390/su12104282

Zimon, D., & Malindžák, D. (2015). Proposal of quality management and technology model supports a subsystem of manufacturing logistics. *LogForum, 13*(1), 19–27.

2 Deviations in Operational Culture
A Barrier Analysis

Salvador Ávila Filho and Beata Mrugalska

CONTENTS

2.1 INTRODUCTION

Understanding the environment (organizational, physical, or cultural) that influences the workplace makes it possible to identify which factors change human behavior to plan preventive actions, improving future scenarios. Organizational climate change always causes changes in the behavior of the operational team, not detected. Managers do not perceive indirect or weak signals relevant to avoid failure and incidents. The perception of deviations requires a calibration of attention in the executive function (Avila and Costa, 2015) indicating the new existing movements of the team in the work routine. These changes affect social, individual, technological, and climate processes through the undue actions of institutions and leaders. In Figure 2.1, there is shown in parallel comparison between workers and ants. It shows what direct and indirect factors can be differentiated that change the task of collecting leaves in the ants' routine. Where does the group of ants (operators) change your routine? Moreover, there arises one more question: how to maintain its efficiency despite the environmental and social noise?

The organization does not detect deviations, which makes it difficult to program barriers to avoid communication noise. The breakdown of the communication chain affects decision and action at various levels, from strategic, managerial, to operational. At the strategic level, the noise created alters the mind map of the original project by top management, indicating a different written pattern. At the managerial level, communication noise makes it difficult for the operator to understand the written standard, leading to an incorrect orientation of the panel

DOI: 10.1201/9781003383444-2

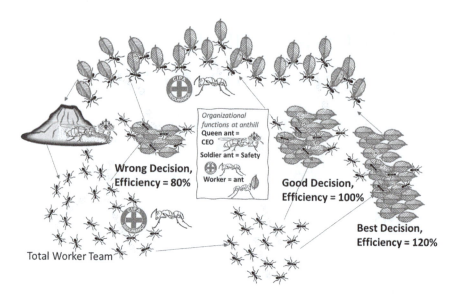

FIGURE 2.1   Ant group decision in routine.

Source: Adapted from Ávila et al. (2022).

to the field. At the operational level, the operator makes a mistake performing his field duties.

The design of the workstation, the development of skills, the group's commitment to specific functions, and the ability to plan actions in complex chemical processes (Ávila, 2012; 2013) can avoid noise and reduce accidents. Indicators related to deviations from routine activity may not be perceptible in the cognitive field, but it is detectable in the intuitive and emotional field. Therefore, the review of indicators depends on assumptions and heuristics of the local culture.

2.2   STANDARD SOCIAL FLEXIBILITY VERSUS TASK DEVIATION

Rasmussen (1997) mentions that there is social flexibility in relation to the standard established in carrying out routine activities. The worker's perception has a natural characteristic arising from life in society, where in the final stages of the task, the standards do not follow completely. Another factor that can influence non-compliance with the standard is the non-availability of optimal resources for the task.

Ávila (2010) confirms in the oil industry that 85% of written procedure is reality, in the field. The rest depends on improvisations. This discussion indicates that perhaps an increase in tolerance can decrease nonconformities. What about high-impact tasks like fluidized catalyzed cracking control? In this investigation, the characteristics of the local social culture may indicate a loss of commitment to the standard, or a loss of quality in working memory. The references poorly established due to the lack of identification of communication signs and cultural symbols need investigation.

This cultural characteristic, added to the underreporting resulting from a culture of guilt, reinforces the formation of a combined memory (Ávila, 2010) based on fantasy. In this way, for different reasons, a tendency toward pattern loss may occur. This phenomenon depends on the construction of appropriate organizational barriers or human elements (Ávila et al., 2019; Ávila et al., Ávila and Pereira, 2020). Some organizational barriers can be built into the company project: establish commitment through a blame culture; develop operational discipline program in genuine format (not related to power); discuss the relaxation of the standard; and analyze the linguistics for procedures (codes, signs, symbols) that facilitate communication for the operational team, both during shift and administrative hours.

A deviation does not cause visible changes in the energy consumption of an industrial plant, or in the logistics for its operation. The physical and cognitive efforts to perform the task are control items within a socio-technical analysis. These requirements demand system redundancies to maintain the normality of quality and quantity produced by an industrial facility or technological service.

Small deviations from routine can indicate the social flexibility discussed by Rasmussen (1997). In this case, it is possible that an amount of hazard energy is released toward the accident, following the Swiss cheese model (Reason, 2003). In a situation of prolonged production, this *hazard* energy quantity reduces. If cultural, human, and organizational factors have a major effect on standards and procedures, the amount of hazard energy released may be greater, raising concern about the high frequency of operational deviations. Finally, including the hypothesis that underreporting is the effect of a global cultural event (Ávila et al., 2021), deviations quickly migrate to variations in process states and task activities facilitating the path to accident, chain reaction, and disaster occurrence.

In this way, we seek to identify when there are deviations that cause individual failures with high resilience, and which are the deviations of low resilience. These deviations can be the result of addictions, bad habits, inadequate short- and long-term memories, or even the repetition of inappropriate rituals (endemic organizational failures) that shorten the path to accident.

The nature of rules and laws, formal or informal, that we establish in routine and unusual tasks, can influence the appearance of deviations from these rules and standards. The development of derived archetypes (Ávila and Menezes, 2015) may indicate this new dynamic risk which can be analyzed for areas of behavior in operations, such as safety, quality control, and environmental impacts.

A rules-only culture does not place value on observing behavior and negotiating acceptable standards within a dynamic environment. Many rules, laws, or requirements can decrease the likelihood of detecting deviations. In addition, the regional, national, and international (global) culture causes biases in the making (in routine or in emergencies). It is important to carry out an analysis of social and human risks in a cross-matrix in the industry control room (Ávila and Pessoa, 2015).

Embrey (2000) applied a survey to operators in the fertilizer and oil industry regarding the main reasons why the operator did not follow the pre-established procedure, and found main aspects related to the style of writing and the complexity of the task (Ávila and Amaro, 2015).

Can communication noise affect the acceptance of deviations? Could the origin of this noise come from difficulties in organizational change? (Ávila et al., 2015). On the other hand, with relations by chronic stress that alters the priority and calibration of attention (Ávila, 2011).

The reasons and consequences for the occurrence or recurrence of deviations in the routine are the number of rules, requirements, and steps in the procedure and what are the memory limits. In a similar mode, the urgency of time and inadequate formation of working memory causes deviances. The memory quality in periods of return from vacations and the relevance and conflict of interests in the discussion about security and business economy cause noises and deviations. The level of complexity of processes and tasks affects communication and causes deviances. In the case of lack of clarity of procedures, lack of definition of priorities and multiple omissions are part of deviation causal nexus.

In the discussion about omissions, we seek to understand the reason for their occurrence from procedures in civil construction where the steps of the task or operations have a high load of information, reducing the operator's immediate memory capacity. Other characteristics that can cause omission are the task steps that are barely visible or imprecise; weak or ambiguous signals (same importance level) initiate steps; the steps are functionally isolated from each other and have no apparent requirements; steps without due attention after reaching the main objective; the steps are repeated and eventually change, not being noticed in duplicate. Task steps modified and not communicated properly due to having operations performed after previous operations interrupted can cause low profit.

The increase in socioeconomic affective utility (Ávila et al., 2022) in the task improves cognitive quality: in the application of established concepts, in the formation of new concepts, and in the decision after the cycle of thinking, deciding, and the organization's feedback.

The organizational culture based on competence, a sense of justice, and clarity of communication is a favorable environment for the nonexistence of deviations as indicated in the recommendations for the sulfuric acid plant (Ávila et al., 2016a) and discussed in the papers of Just Culture (Bitar et al., 2018; Amalia, 2019; Pellegrino, 2019).

The areas of people selection, skills development, routine operational control, stress control for better decision quality, and understanding cultural biases are extremely important to reduce operational deviations. Keeping characteristics such as cooperation and commitment at a high level improves the group's vision and intervention in operational control.

The construction of the task depends on the analysis of the work. The view of operational standards and procedures (Ávila et al., 2022) with the analysis of the environment, cognitive processing, task control, and task failure analysis is necessary to avoid slips due to lack of knowledge regarding the tasks performed.

Barriers may have low efficiency, with failures in the training or selection of people, so we must take care of the follow-up and monitoring of cognitive and organizational barriers. Excessive reliance on technical barriers and underinvestment in cognitive and social barriers promote deviation to failure and promote failure to accident. These barriers improved from the risk analysis and the recognition of the cause of deviations in the design and operation stage can construct an excellent risk management program.

The communication style seeks to avoid the accident around safety and seeks to increase the efficiency of the task. It is important to measure the efficiencies in the various types of communication based on Pyramids Model (Drigo and Ávila, 2016). Different dimensions of communication need evaluation, from written information about safety including hazard mechanisms at the workplace until the review and response to workplace hazard. Written operating procedures, training material (safety and operation), and information flow are important for the contractor and people. The emergency response training is an important dimension, principally when included in a change management moment. The emergency response plan for the public has specific characteristics for effective scape in the disaster.

The current form presented by documents does not allow the investigation of complex processes (Perrow, 1984) and the investigation of tasks and complex social relationships (Ávila, 2013). We should always include the discussion about the effects of increased stress on decision-making and worsening of the process safety culture (Ávila, 2011; Mrugalska and Dovramadjiev, 2022). We seek to install an Integrated Management System focused on operational.

2.3   INVESTIGATION OF DEVIATION NORMALIZATION

Cultural biases that become vices can take time to saturate an organizational and technological system, or they saturate quickly. It depends on the robustness of the organizational and safety culture, promoting, or not, events that can lead to the accident. The undertaken decisions have a variable tolerance regarding the level of risk and depend on the management's perception of risk.

Employees of an organization may repeat deviations that are the result of primary or secondary archetypes. Workers can also adopt bad habits being more subtle than deviations, but sometimes it brings worse consequences for not having visibility. These deviations can change over time, due to the change in the characteristics of the hazard, they end up changing the probability of the accident happening. Therefore, culture-related methods predict a dynamic situation with the possibility of changing behavior.

The frequency of deviations increases with (1) the lack of knowledge of the danger; (2) by not believing in event risk; (3) by underestimating the severity of the event; and (4) in the event of cognitive disconnection. The Executive Function Model (Avila and Costa, 2015) indicates characteristics to perform the task in planning with hot and cold components, where subjectivity inserted in the warm components is intuitive and emotional; and objectivity embedded in cold components is cognitive. Understanding the heuristic of decisions, based on the cultural matrix, and the respective meanings allows a quick decision through a mental map, in this case a working memory with immediate rescue. The analysis of the mind map depends on the level of urgency in the decision, indicating a dynamic format regarding the level of stress that can be progressive (Ávila, 2011).

The reason to discuss the phenomenon of normalization of deviations is due to the uncertainty of human error in the planning-execution of the task and in the actions of prevention, correction, and mitigation to avoid accidents. The relationships between deviations, incidents, and accidents can lead to a catastrophic event if there is a chain reaction of the action of unknown hazards. However, it can also lead to a near miss that, untreated, causes disaster. Although human error in task execution is identifiable, human error in leadership or management may not be so easy to map due to the relationships between the power of organizations and the role of investigating failures. The organizational, natural, and social environments influence the relationships installed in the workplace, and in particular the horizontal–vertical communication processes and the flow of information (Drigo and Ávila, 2016).

The Safety Culture Investigation indicates a new graphic format that is no longer the accident pyramid. We can say that it turns into a Christmas tree with regions of underreporting of events before the occurrence of the accident with lost time or even before the chain reaction to disaster.

Human error is a symptom that indicates problems in the organizational environment. Human error is not the root cause of the accident. Training and task review actions do not resolve or prevent the causes of the accident. Complex technological systems, for example, the case of air transport control, and when these systems cause a high impact in the event of an accident, redundant barriers installed from the deviance stage. The failure of process safety is a symptom that indicates problems in the safety culture and/or organizational culture. The installation of barriers demands knowledge about the relationship between technology, barrier design, selection, and control of people.

The lack of operational discipline, the low tolerance for deviations in operational practice, and the complacency in relation to risk situations are important features in assessment. The Bhopal incident is a lesson learned that used repeatedly to indicate that organizational decisions promoting lack of maintenance causes leakage and hazard release situations. Conditions deteriorated over time, with deviations and despite the flaws being visible.

People design wrong solutions to avoid accident or disaster because they do not know exactly what the nature of failure is. Unknown how to discuss where the various deviations and bad habits occurred, and which trigger set off the chain reaction. Everyone invests in a technological cause or human error indicating problems of misbehavior at work.

The men/women at work are not able to observe and respond, because, they cannot build working memory based on reality, in this way the information for decision is incomplete. There are difficulties scheduling or rescheduling the task, again, there are problems in revising the original instruction. If you cannot work with resilience and engineering, so the worker cannot improve processes.

Deviation caused by not knowing how to manage technological and organizational changes is noted. It is a case of reviewing the security requirements during the pre-start procedure. A program that considers cultural aspects in the routine aims to achieve operational excellence.

The relationships of the types of deviation: managed deviation, where previous actions seek to avoid its occurrence; apparent manageable deviation where actions are not located in the root cause; and the occurrence of unintentional deviation, and the difficulty of identifying the deviation in their work process.

An accident that leads to the coma of an industry worker, in a real case of onshore oil production, where the worker, instrumentalist, arriving at the workstation, seeks a technical document to carry out maintenance. Due to a difficult and restless journey, not complying with the proper ritual for the service station, the instrumentalist retrieves the wrong document, which does not indicate the existence of a spring. The incorrect ritual of adaptation between the external environment and the workstation, where immediately after disembarkation, maintenance began, caused the valve to open with incorrect design, opening as if it were without a spring and had a spring, thus the serious accident occurred.

Organizational communication (Drigo and Ávila, 2016) influences the existence of deviations and accidents from failure in communicative processes. Some defects in applied linguistics and inconsistent information flow in its properties, intensity, object, objective, direction, and priority can influence in deviance trigger. The priority standards, occurrence records, and feedback are important parts to be analyzed avoiding deviation. Other characteristics are assertiveness and goal in the guidelines and requirements for the work.

The policies in the organization to handle routine and emergency are important in the routine, but the communication defects during routine and contingency. The types of communication indicate circulating information upward, horizontally, and downward according to the hazard's mechanism of action. A question, does low quality in communication promote the repetition of the deviation?

Knowledge, skill, organizational relationships in employees' psychological and legal contracts with companies do not guarantee the establishment of appropriate commitment. A series of predicted or unforeseen cultural events can induce variation in human behavior and performance factors. The cultural events can induce behavior variation that be noticed by the leadership for the future activities.

Living with deviations without proper feedback on corrective needs and their prioritization creates the organizational environment for process losses. The omissions that occur in the routine and the lack of feeling of ownership for the task are characteristics that enhanced in the shift regime where diverse and adverse situations occur. During shift hours, in a specific organizational culture (communication), the written procedure may not be valued (short time is applied in its elaboration) and horizontal communication may not be taking place properly on the shop floor!

Understanding the importance of informal rules (heuristics for decision) and procedures on the shop floor is a condition for engineers and managers to know how to deal with cultural biases in the routine of the operation. Establishing redundancy in actions in advance avoid decision errors. It is important to carry out the risk analysis from the reading of the operator's speech where hypotheses about the chain of abnormalities are discussed, where many informal rules format the operational routine.

Knowledge of formal work rules and critical tasks are necessary but not sufficient to avoid deviations, it is necessary to understand the operational culture.

The operational culture (Ávila et al., 2022; Ávila, 2010) established in industrial plants and in the routine of their operations is the result of a field of cultural forces. We can say that one or two of the characteristics in this force field may be dominant. Thus, operational culture is the result of (1) influence of regional and global culture; (2) intrinsic personal characteristics in key functions in the flow of operations; (3) influence (positive and negative) of the immediate and mediate leader; (4) organizational influence (positive and negative); and (5) common sense as a citizen and family member.

The Workplace Project structured to meet the characteristics of technology and socio-organizational systems and their changes. In order to design the job and build the appropriate team for each objective, considering the current dynamism, we must identify several properties: cultural biases and defects in feedback, and flaws in the operator's socio-functional environment at his job and its related risks.

Other properties are part of discussion, then, the minimum requirements for cognitive functions in the worker's life cycle, organizational movements, and the role scheduled and performed in the operation activity compared with performance factors.

The types of human error, cognitive gap/failure, and relationships with process loss and accident are part of deviance root causes. The risk of changes due to unavailability of facilities and resources can show possible failures. Finally, we need new indicators for cognitive deviations and failures, and a sketch of the failure in complex mode points connecting human errors and technical failures, task cycle, noise, and failure body. What are the hazard mechanisms involved in this social phenomenon?

The Management of Technological Change (MOC) and Organizational Change (MOC) are difficult times for a high-risk activity (process, product, and task) with the possibility of damage to property, workers, neighbors, and the community. In this activity, it is important to understand the appropriate standard, have experience in this activity, and the discipline to ensure the expected quality, quantity, and sustainability.

The activity carried out respecting the limitations of psychological functions and after analyzing, the rituals within the factory can format good barriers. From this study, during the elaboration of the task, we seek to adapt new rituals to adjust archetypes and eliminate bad habits (Ávila and Menezes, 2015). These questions travel in the cultural dimension with identification and measurement difficult to carry out.

In the case of emergency that require behavior changes, the need is even greater to discover the appropriate leadership profile to mitigate these situations and reduce their impacts, the emergency leader has the need for command and speed in actions. Within the work team, we must provide support to address or treat the danger released from the chain reaction.

The transient process migrates from one state of the industrial process to another; it can be at the start of the plant, or in the change of load, or in the change of processing materials in new production campaigns. A necessity: set the default states for each production campaign and set the target state.

In terms of design, these definitions are part of the technological packages, but establishing what is normal or abnormal when the plant is operating depends on the availability of people, equipment, and the suitability of raw materials. Great care about the incorrect behavior does not accept transient process conditions or consider transient as normal conditions. It repeats in terms of operation, what constitutes a deviation, when the deviation is part of the task, and when the deviation occurs in a random and harmful way for the plant. Operations team leaders and operators with specific roles should have the ability and authority to stop the process in the event of a high-impact hazard release. Therefore, leaders and operators must understand cognitive failures and their relationship with human performance factors and process losses. This understanding of causality allows us to understand the risks of decisions during change.

One way to avoid plant downtime, production delays, and accidents during a technology change (new product, operating range and limits, increased load, and task changes) is to do simulated testing and risk analysis. Another widely used technique is the study of equipment failure modes (Ávila et al., 2012), which should include the analysis of procedures and performance factors for operators and maintenance technicians. The relationship between the task, the man, and critical components indicates the importance of detecting the cognitive failures of these workers. An analysis of FMEA includes human factors in the investigation of insistent plant shutdowns caused by hydrogen turbocharger (Ávila et al., 2013). There are many possibilities to inhibit deviations.

The major difficulty in failure and risk analysis is that incremental changes over a long period may not be detectable because attention focused on issues cause immediate economic loss. Small deviations are not valued, the routine occurrences recorded during the shift should feed the system that processes the data and indicates the tendency for failure.

In many moments, the lack of knowledge can generate a situation of insecurity and the pressure of time induces the acceptance of deviations. The lack of knowledge caused by improper selection, improper training, or not having a culture of lessons learned in operational practice indicate possibilities of deviance. Undue knowledge or skill can affect the operation's work; it can generate a bank of inappropriate knowledge promoting an improper situation in the operation. The intention to train and provide the necessary knowledge may not translate into action due to strategy errors and inadequate acquired facilities.

Cognitive gaps caused by lack of knowledge can lead to inappropriate decisions (Avila and Costa, 2015). Unknown mechanisms of release of known (or unknown) hazard in chemical processes can cause incident or accident. Normally, the trainings do not deal with the hazards release from operational deviations. On the other hand, even when they happen in an appropriate format. The training discussion on lessons learned adopts an improper format that believe totally in the super-operator, in the BIGDOG, when in fact it is not.

Collins (2001) states that the process safety leader who knows how to avoid disasters during the chain reaction should be extremely humble and practice full-time active listening with strong positions and fierce determination in actions until the hazard line needs identification. It does not aim at power in organizations, always cooperates, but demands attitude in urgent need. Thus, considering yourself the most knowledgeable of situations, or as BIGDOG claims, is not a Process Safety Leader's strategy.

The discussion about what are correct knowledge and skills requires surveys that cannot be superficial and therefore require confirmation of the reasons for human errors and technical failures through a historical analysis of events and operational routine. Thus, we seek to compare the competence offered (analysis of past human errors and examination of the operational culture for decision) with the competence demanded (technology and operation characteristics) and define a more appropriate training program (Ávila, 2010). On the other hand, we can analyze the operators' perception of the risks of routine operations based on surveys and compare with the suggestion by the leadership (Ávila and Santos, 1999). In neither of the two tools, we affirmed that the greatest knowledge belongs to the oldest supervisor, the one nicknamed BIGDOG.

Lessons learned from incidents help staff to analyze root causes, needing good records. These lessons transformed into training material and new procedures show through similar sceneries the facts lived. Post-treatment incident records generate training material that reviewed for inherent dynamic risk characteristics in the process. Understanding the practice of a scenario is not enough to guarantee the non-recurrence of accidents. Open social environments can alter behavior and the accident will occur in different ways, or will manage to bypass barriers through primary failures, but derived from culture. Thus, it is necessary to anticipate the direction of the accident through dynamic risk analysis that introduces the vision of environmental influence.

Retirement and transfer of knowledge to the younger ones, the question of the clash of generations in attitudes in shift work, is where the command versus argument dilemma occurs! Moreover, other social conflicts such as "virtuality" and the clash of generations indicate the need to review tools in the chemical process industry and in other risky activities. What will be the next social conflict?

Does the sustainable organization have intense changes in leadership or few changes? What is the importance of social relationships at work? Why might each new person be inadvertently introducing detours into work? The chemical companies encourage skill-based multitasking sustainability.

Restrictions on resources lead to cuts in people and materials, cuts made without criteria, leading to a reduction in the application of resources for the maintenance of safety barriers, reduction of people, and leadership. Not working correcting the tendency for the accident leads to situations like Piper Alfa and Texas City.

Human performance factors are closer to job characteristics that authorize interaction between worker and equipment. These human factors are part of the structure and functions of the flow of information that circulates in the workplace. Being able to be closer to man bringing the possibility of failure in information processing. Being able to be closer to the equipment where technical knowledge requires controls and actions at the right time. The human–machine interface has a high information flow density and high complexity. The regional or global characteristics affect the relationships in the workplace, making it difficult or facilitating the accomplishment of the task. In the AHFE, a matrix proposed that involves cultural characteristics that influence and determine the possibility of cognitive failure in certain constituent parts of the panel-control room of LNG (Ávila and Pessoa, 2015).

Do we have the internal skills and work capabilities necessary to instruct the work without affecting the performance of day-to-day tasks? According to Muchinsky (2004), the expectation of competence when planning a training for the team does not coincide with the competence developed after the training period. Why? Because we may not audit the quality of work after training and the level of adherence of new training to work processes.

Employees cannot perceive the leadership message as being "who can command and whoever has the sense obeys". Alternatively, even "take your leaps and solve your problem". Thus, care is necessary in the discussion about punishment and reward to achieve production goals, to be innovative, to save energy and materials. Punishment and reward actions can stimulate anomalous forms of behavior in organizational cultures. The prizes offered in the past to avoid deviations may encourage new deviations. Does the organization accept deviations?

Catastrophic events are not accepted by the team to risk investigation. The team, within the work facility, does not accept the view of risk about some rare event. This makes it difficult to project a future accident: become difficult to build an appropriate safeguard from the lessons learned and understanding new deviations can occur.

An experience in the LPG industry has shown that to build a future accident, one must investigate past deviations and events and jointly build this new accident direction with the staff. Operational staff resist considering this possibility because of the blame for being responsible for a major accident.

Addictions created from repeated forms accepted by the group, from a very old regional, local situation, present in the collective unconscious or even at a primary level, contribute to accident occurrence. Some vices come from misbehavior created in the organization when the shift behavior is separated from the administrative one indicated by archetypal behaviors.

The management system is sick when formal reports are more important instead of analyzing routine facts, as they are. Occupational safety cannot be part of a pure inspection effort, but rather, learning new values or practices around safety. The organization of the work indicates a great concern of the operational functions in the management tools with excessive time of the maintenance and operation team in the elaboration of reports, in the participation in meetings, and in the audits in excess. Today we are losing the skill-based competencies, which are precisely those needed to recognize the reasons why deviations cause accidents.

Safety training for contractors must present the real dangers and risks by experienced instructors without which accidents will continue to happen. There is still the possibility that, due to the economic bias of the minimum price, the contractors acquire leadership and systems that are not appropriate in practice, although they meet the requirements of the integrated management system. It meets with immense difficulties, and then we conclude that the accident is certain! The risk of an accident increases when considering the process of issuing a work permit, where the causal nexus not investigated decrease reliability of the services. The focus is on filling out endless checklists, no time to investigate problems. The operation team do not have ability to notice changes and deviations that can indicate a serious change of state.

Change management should include the analysis of complexity in conjunction with the classic risk view that discusses frequency and severity. Complexity analyzed at the technology, task, and communications levels, through a socio-technical risk analysis, indicates the ways in direction to accident. The investigation of accidents that have occurred and the projection of future accidents done by discretizing the operator's speech and joint analysis with process and production variables is possible to identify operational culture. On the other hand, the emergency plan, inserted in the management of changes, must have a command leader and his team, having as main characteristic the quick cognitive decision (Ávila et al., 2015).

Excessive self-confidence can be the main cause for the phenomenon of normalization of deviations. In this situation, we need to understand the current personalities in the operation team. The difficulties of managing multicultural, gender, and generational conflicts make the BIGDOG question themselves about the need for argumentation since their experience would be enough for the command action. Thus, BIGDOG is self-confident and enforces its decisions. On the other hand, unskilled people may suffer from the illusion of superiority, mistakenly believing that their judgment is exceptional!

Decisions based on lessons learned can direct barriers to inappropriate positions where excellent past performance ends up bringing the comfort of an unrealizable future scenario. The routine team does not want to analyze deviations to build a future accident, they are afraid to discuss events that could result in an accident. It is easier to talk about accidents that have occurred than environments that could lead to a future accident. It does not want to suggest routes and directions for the path of danger until the accident takes place.

The lack of knowledge about cognitive gaps leads to the need for excessive redundancy in safety instrumented systems indicating that when automation fails, I cannot think of solutions that I am not trained in. Are we justifying deviations with an open story? No respective causal link.

Overconfidence in technology and simulations indicates the lack of human reliability in skills, and also the need to seek balance. Changes in the workplace can stress teams and leaders. SIS can generate the phenomenon of cognitive laziness.

The disease of integrated management discusses about excessive flow of information and a lack of field skills. It feeds the phenomenon of "virtuality" and the installation of a new generation that is more dynamic and visionary, but linked in virtual world, and less applied to field practices (Ávila et al., 2016b).

Is there a limit to automation as a process performance tactic? The balance between complex decisions based on the study of field signals with the interpretation of scientific models that indicate how to control processes. When will we learn to manage dynamic risk?

The Brazilian culture and part of the global culture seek leadership to follow, or even, not having established leadership and not assuming the same, adopts the behavior of excessive flexibility and avoiding conflict in decisions. These aspects of human nature but based on social culture join the movement of egocentrism where being a boss or owner is more important than cooperating, bringing an addictive trait of paternalism.

Actually, genuine leadership values are overridden by intended organizational values, resulting in conflict and low sustainability. Thus, companies that are built assuming the coexistence of differences must understand the phenomena of socialization (mediated by a system of reward and punishment), institutionalization (authorities present as norms), and rationalization (enable operators to convince that their deviations are legitimate, acceptable, and necessary) allow corrupt practices to evolve in white-collar organizations.

New concepts already presented by Ávila discussed in this work clarify the origins of deviance. Organizational jelly discusses the responsibility for supervision influenced or shared by informal leaders. Other models treat about conjugated memory—working and long-term memory undergo protections and transfers indicating that the constructed scenario does not exactly represent physical reality.

The cognitive process with perception and information processing is based not only on cognitive aspects, but also on affective (psychosomatic) and intuitive relationships, making it more difficult to perceive real risks in complex systems. The decision by priority orbitals treats the understanding about dynamic decisions that depend on the pulse of energy coming from the resolution of the internal, external world and the workstation.

The good decision is resultant from information flow and choice of alternatives as to how to act in the activity, business, or institution. When the information comes from the internal metabolism of the tasks, it is important to know how to perceive the deviations, including the unsaid ones. When the information comes from external movements of those interested in the company, externalities, it is important to transform it into an internal language to understand. Internal information depends on feedback that depends on the comfort of the safety culture. Thus, the application of the punishment–recognition strategy reviewed in the direction of a learning culture, without which there will be underreporting and difficulties in decision-making. At the same time, perceiving behavioral changes using operator discourse analysis as a practice allows reviewing risk management (Avila and Drigo, 2015).

A certain shift leader asked me what to do when his subordinates do not respond immediately, when actions are demanded. There are many questions about the right moment and the reason for such an action, so this leader based on a command style does not know how to lead in the argument style, there will be difficulties for decision and action with the conflicts of generations.

Poor quality communication results in a low valuation of the perception of risk, in fact, and brings the blame of the applicant organization in relation to the accident. A low work value is assigned to the sincere perception of risk based on facts, and a transfer to organizational protection created in non-real and non-guilty scenarios shortly.

2.4   DEVIATION ORIGINS

The rescue of the debate initiates the discussion of the origins of the deviation. Initial learning about cooperative processes reduces communication noise and information losses on the way to work. The pattern of work may change from its actual requirements for carrying out the task and transforming the economy. The dynamic requirements for the task can be of technological, social, or organizational origin, passing through the movement of information from a formal to an informal level in the Operational Culture dimension. This event directly affects the planning, execution, and review of the task, through fault mapping and the need for technological changes. Then comes the questioning of which factors increase the probability of deviation and which are the possible interventions in the operational routine to avoid human error.

This subject is very subjective and has high capillarity, where environments and factors change reality quickly. In addition, this subject is difficult to access; the flow of information available does not reflect a scenario closer to a pure reality. A high level of uncertainty sets in, despite the feeling of working between appropriate ranges of process variables, the standards present a false reality of security, we fail to identify the technological, cultural, and task complexity. The management tools of Risk and Failure Analysis are not enough to detect cultural problems on the "shop floor" and at that moment, we try to discuss human and social factors together with the technological ones in an organizational environment.

In this situation, the meeting between specialists who have different views intersect in a socio-technical scenario, with great knowledge and skill in the routine of the chemical process industry, where the risks of accidents related to products and processes linked with cognitive processing. Those experts who open a discussion without preconceptions agree that cultural phenomena at work can alter human behavior and align channels of migration of the energy of danger toward the accident.

The "experienced" Process Safety Engineer maintains deep level discussions with the Cognitive Processing Engineer at Global Congresses and specific meetings opening the dimensions to study how the acceptance of deviations in the industry routine occurs.

In Figure 2.2, we present the opening of the discussion on the variability of behavior that affects Industry Safety through the dynamics in Human Factors.

For this purpose, the Process Safety specialist presents the following focal topics for the discovery of this new dimension.

In the Behavior dimension, we seek to understand how to compensate for the intrinsic variability of human behavior. In the Safety dimension, we try to understand the low effectiveness of current techniques offered in the classic view where the accident has only technological origin. In the Information dimension, where the routine flow, quality, and perception of signals transform practice into knowledge that is the basis of operational decisions. In the Lessons Learned dimension, we learn about professionals from defense agencies, military, police, and firefighters,

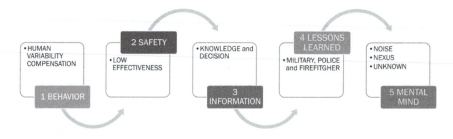

FIGURE 2.2 Human factors variability.

with operational discipline faithful to the risk of the process, thus demonstrating the loyalty of purpose. In the Mind Map dimension, the noises that affect the reality scenario, the combined memory that approaches or moves away from the real memory, and the difficulty of establishing a causal nexus about processes that involve unknown danger are discussed.

2.5 CONCLUSIONS

Culture is perceived as a learned behavior which is not a side effect of operations or an overlay. It is created in each organization due to undertaken actions. Moreover, it influences the employees' work behaviors in a form of acceptable behaviors and attitudes. Therefore, understanding the work environment (organizational, physical or cultural) allows identifying factors which change human behavior to be able to plan preventive actions and improve future scenarios. On the other hand, operational culture can be both a priority and a challenge as it is difficult to precisely articulate and deal with culture. Organizational climate change always leads to variations in the performance of the operational team which are not detected. The organization is not able to identify deviations that lead to the problems with programing barriers to avoid communication noise. However, there is social flexibility in relations to the standards which refer to routine activities. But in practice a tendency toward pattern loss can appear due to not well-established or lack of organizational barriers or human elements. In this chapter, it was shown that the frequency of deviations increases due to such factors as lack of knowledge of the danger, not believing in event risk, underestimation of the severity of the event, and in the event of cognitive disconnection. The problem of normalization of deviations and their origins was widely presented as it refers to the uncertainty of human error in the planning–execution of the task and the actions of prevention, correction, and mitigation to avoid accidents. All these discussions lead to opening a discussion and further research directions toward the variability of behavior which influences industry safety through the dynamics in human factors.

REFERENCES

Amalia, D. (2019). Promoting just culture for enhancing safety culture in aerodrome airside operation. *International Journal of Scientific & Technology Research*, 8(10), 260–266.

Ávila, S., Santos, A.L.A. (1999). *Environmental suitability with clean routines in the chemical industry*. Ecolatina'99, 2nd Latin American Conference on the Environment, Belo Horizonte, 12.

Ávila, S. (2010). *Etiology of operational abnormalities in the industry: Modeling for learning*. Thesis (Doctor's degree in Chemical and Biochemical Process Technology), Federal University of Rio de Janeiro, School of Chemistry, Rio de Janeiro, 296, 2010.

Ávila, S. (2011). *Dependent layer of operation decision analyzes (LODA) to calculate human factor, a simulated case with PLG*. 7th Global Congress on Process Safety, GCPS, Chicago, 21, 2011. www.aiche.org/conferences/aiche-spring-meeting-and-global-congress-on-process-safety/2011/proceeding/paper/81s-dependent-layer-operation-decision-analyzes-loda-calculate-human-factor-simulated-case-glp-event-0.

Avila, S.F. (2012). Failure analysis in complex processes. In: Proceedings of 19th Brazilian Chemical Engineering Congress—COBEQ, Búzios, Rio de Janeiro.

Ávila, S.F. (2013). Failure analysis in complex chemical processes. *Industrial Chemistry Magazine Revista Química Industrial XXIII Year*, 139, 24–25. ISSN 0103-2836.

Avila, S.F., Costa, C. (2015). *Analysis of cognitive deficit in routine task, as a strategy to reduce accidents and industrial increase production.* In: Safety and Reliability of Complex Engineered Systems, London, pp. 2837–2844.

Ávila, F.S., Amaro, R. (2015). *Usability of procedures based on human factors assessment, a case of petrochemical industry in Brazil.* 6th International Conference on Applied Human Factors and Ergonomics (AHFE 2015), Proceedings, Las Vegas.

Ávila, F.S., Menezes, M.L.A. (2015). *Influence of local archetypes on the operability & usability of instruments in control rooms.* Proceedings of European Safety and Reliability Conference—ESREL, Zurich.

Ávila, S., Drigo, E. (2015) *Operator discourse analysis as a tool for risk management.* European Safety and Reliability Conference, ESREL, Zurich.

Ávila, S., Dantas, E., Duarte, J. (2015). *The requirements & tools to treat process safety risk as result of organizational change.* 11th Global Congress on Process Safety, GCPS, Austin, 2015. ttps://aiche.confex.com/aiche/s15/webprogram/Paper396803.html.

Ávila, S.F., Pessoa, F.L.P. (2015). *Proposition of review in EEMUA 201 & ISO standard 11064 based on cultural aspects in labor team, LNG case.* 6th International Conference on Applied Human Factors and Ergonomics (AHFE): Procedia Manufacturing, Las Vegas, 6101–6108.

Ávila, S., Santino, C., Santos, A.L.A. (2016a). *Analysis of cognitive gaps: Training program in the sulfuric acid plant.* Proceedings of European Safety & Reliability Conference—ESREL, Glasgow.

Avila, S.F., Fonseca, E.S., Bittencourt, E. (2016b). Analyses of cultural accidents: a discussion of the geopolitical migration. In: Proceedings of Annual European Safety and Reliability Conference (ESREL), CRCPRESS, Glasgow.

Ávila, S., Mrugalska, B., Ahumada, C., Ávila, J. (2019). *Relationship between human-managerial and social-organizational factors for industry safeguards project: Dynamic Bayesian Networks.* Proceedings of the 22nd Annual International Symposium Mary Kay O'Connor Process Safety Center, College Station, TX, 22–24.

Ávila, S., Ávila, J., Pereira, L.M. (2020). *Reviewing tools to prevent accidents by investigation of human factor dynamic networks.* International Conference on Applied Human Factors and Ergonomics. Springer, Cham, 233–240.

Ávila, S., Souza, L.F.L., Costa, G.J., Pereira, L.M. (2021). *Black swan team: Finding competence gaps.* Virtual 17th Global Congress on Process Safety, GCPS. www.aiche.org/academy/conferences/aiche-spring-meeting-and-global-congress-on-process-safety/2021/proceeding/paper/117bu-black-swan-team-finding-competence-gaps.

Ávila, S., Santino, C., Cerqueira, I. (2022). *Human factor and reliability analysis to prevent losses in industrial processes: An operational culture perspective.* Elsevier, Amsterdam, 1st edition.

Ávila, S.F., Mendes, P.C.F., Carvalho, V.S., Amaral J., Mendes, P.C.F. (2012). *Human factor analysis in equipment failure FMEAH of turbo-compressor at chemical fields.* 27th Maintenance Brazilian Congress, Promotion by Abraman Rio de Janeiro, Proceeding of CBM 2012, Rio de Janeiro.

Bitar, F. K., Jones, D. C. Nazaruk, M. Boodhai, C. (2018). *From individual behaviour to system weaknesses: The re-design of the just culture process in an international energy company. A case study.* Journal of Loss Prevention in the Process Industries, 55, 267–282.

Collins, J. (2001). Level 5 leadership, the triumph of humility and fierce resolve. *Harvard Business Review*, 79(1), 67–76.

Drigo, E.S., Ávila, S. (2016). *Organizational communication: Discussion of pyramid model application in shift records.* AHFE 2016 Applied Human Factors and Ergonomics Conference, Walt Disney World Dolphin & Swan, Florida. Advances in Human Factors, Business Management, Training and Education, 739–750, July 27–31.

Embrey, D. (2000). *Preventing human error: Developing a consensus led safety culture based on best practice*. Human Reliability Associates Ltd., London, 14.

Mrugalska, B., Dovramadjiev, T. (2022). A Human factors perspective on safety culture. *Human Systems Management*, 1–6 (in press).

Muchinsky, P.M. (2004). *Organizational psychology*. Pioneira Thomson Learning, São Paulo, 7th edition, 508.

Pellegrino, F. (2019). *The just culture principles in aviation law*. Springer International Publishing, Cham.

Perrow, C. (1984). Normal accidents: Living with high-risk technologies. Basic Books, New York.

Rasmussen, J. (1997). *Risk management in a dynamic society: A modeling problem*. Elsevier Safety Science, London, vol. 27, 183–213.

Reason, J. (2003). *Human error*. Cambridge University Press, Cambridge.

3 Pupillary Reaction as a Tool to Control the Psychophysical State of Workers

Marina Boronenko, Oksana Isaeva,
Yuri Boronenko, Maria Kozhanova,
and Anton Kozhanov

CONTENTS

3.1   INTRODUCTION

Today, companies strive to get high-level specialists who not only have the necessary competencies, but also know how to build harmonious relationships with the team. The issue of staffing with highly qualified employees is also relevant for the Ministry of Emergency Situations. It is clear that not all people are able to withstand such working conditions in which employees of the Ministry of Emergency Situations are forced to be. Therefore, in order to prevent injury/death of workers, they initially try to hire only psycho-emotionally stable people (Kuznetsova, 2017). However, despite the existing methods for selecting candidates for a position, in 2021 alone, 30 cases of suicide by employees of the Ministry of Emergency Situations were recorded. At the same time, the pre-suicidal state in 83% (25) cases is accompanied by psychological stress. In addition to standard reasons, such as difficult family relationships,

DOI: 10.1201/9781003383444-3

intrapersonal problems, and employees of the Ministry of Emergency Situations are subject to additional stress factors. Their combination and duration of exposure leads to a cumulative effect. Also, in connection with the digital transformation of society, the selection of candidates for the position remotely is increasingly taking place. Therefore, it is important to find a way to digitally quantify the additional metrics needed to reduce the likelihood of a bad hire. The fact that suicides take place in the system of the Ministry of Emergency Situations indicates the imperfection of the methods and means of controlling the psychological health of employees and job candidates. Thus, it is important to improve the methodology for assessing and monitoring the psycho-emotional state of employees. The goal is to test pupillary response monitoring as a tool for controlling/monitoring the psychophysical state of workers. The results of the research expand the possibilities of online recruiting and increase the likelihood of early detection of a psychologically unstable state of the individual.

3.2   MODERN METHODS OF PERSONNEL SELECTION

Strategies and features of recruitment in each country are different, which is determined by the prevailing national business culture and mentality of a particular country. In the foreign practice of personnel selection, there are two main approaches: American (Western) and Japanese (Eastern) (Kotlyachkov, 2017).

The American approach involves the use of traditional principles of personnel selection in hiring; the focus is on specialized knowledge and professional skills. General criteria for recruitment are education, practical work experience, psychological compatibility, and ability to work in a team.

The Japanese approach implies as criteria: the ability to combine professions, work in a team, understand the importance of one's work for a common cause, the ability to solve production problems, and coordinate the solution of various tasks (Kotlyachkov, 2017).

Conventionally, the selection of candidates for a position can be divided into traditional (the most frequently used), non-traditional (allow determining the type of a particular candidate with higher accuracy), and innovative methods (automation and dissemination of artificial intelligence).

The traditional ones include questionnaires, testing, and interviews. The advantage of these methods is their low cost, in terms of both time and money. During the interview, the candidate can only say the information that he wants to share, and he can also give incorrect or exaggerated information about himself and his capabilities. In the process of completing a survey or testing, a potential employee may give socially expected answers, rather than truthful ones. Also, tests are mainly aimed at identifying weaknesses, not strengths. Therefore, traditional methods cannot guarantee high accuracy of the correct choice of an employee for further work. An analysis of the available methods for selecting personnel for the MES system showed that the MES uses standard (traditional) methods that allow not only determining personal qualities, psycho-emotional state both in general and after exposure to extreme factors (Shlenkov, 2005), but also monitoring the psychological state of employees in order to identify the presence and severity of psychological consequences associated with the characteristics of the activity performed. However, as already mentioned, the standard methods for checking the compliance of employees/applicants with the stated requirements have a number of disadvantages: obviousness for the subject of the direction of the questions, which often disrupts the reaction to the survey in the form of a perceived distortion of answers in cases where a person has a need to look in the eyes of the experimenter as a certain positive manner. Since the possibility of detecting and correcting such distortions in the mentioned tests is not provided, the reliability of the results is low. Second, such scales do not allow for dynamic observation of the subject, since the answers are easily remembered during the first study and are often automatically reproduced during repeated surveys, which distorts the results of repeated testing (Yuzhakov, Avdeeva, Nguyen, 2015). Online recruiting of workers uses these methods.

Non-traditional methods include case methods, special interviews (stress and brainteaser interviews), the use of various sciences (graphology, numerology, socionics, phrenology, physiognomy), and selection by competencies. The use of these methods is quite time-consuming and requires the presence of a special employee who will decipher the data obtained during the selection. With the right approach, online recruiting is quite capable of applying this method.

Innovative methods include polygraph and eye-tracking. But even a polygraph can be deceived with special training. Unlike a polygraph, eye-tracking has a number of advantages. On the one hand, eye-tracking is a less stressful procedure, since the registration of oculomotor activity is carried out without contact, without attaching sensors to the body of the subject, which makes the assessment procedure more comfortable for the subject. On the other hand, the process of registering the position and movement of a person's gaze takes three times less time than a standard study using a polygraph (Bukhtiyarov et al., 2019).

iMotions* offers the iMotions on-screen eye-tracking module. The module allows for advanced analysis using tools such as heat maps, gaze repetitions, and area of interest (AOI) outputs such as time to first fix and elapsed time. The module has the following main features: complete presentation of incentives; integration and synchronization of all types of sensors (eye-tracking, facial expression analysis, electrodermal activity aka GSR, EEG, ECG, EMG), the presence of a built-in survey tool to add questionnaires to the data set. The iMotions eye-tracking module is based on tracking a face tracker or an eye tracker in the case of using the iMotions Eye Tracking Glasses Module, which makes it possible to take the necessary facial expression, which will subsequently allow you to deceive the system. Therefore, we suggest using the pupil as a pattern of emotions/stress state, since in order to "imitate" a change in the size of the pupil, you need to try hard.

For video recording of eye movements, systems based on eye registration with a high-speed video camera operating in the near infrared (IR) range (850–950 nm) are used. The disadvantages of using the method with an infrared camera are the following problems. Based on the fact that there are the following interferences that make it difficult to detect the pupil: painted eyelashes, elements of a spectacle frame that fall into the field of view of the recording video camera, any additional bright elements that fall into the frame: reflections from the surface of glasses and contact lenses; spectacle frame elements; illumination from abnormal light sources, etc.,

---

* https://imotions.com

when conducting mass eye-tracking studies, the optimal strategy is to avoid using subjects whose individual characteristics make it difficult to record eye movements. In practice, compensation for such interference can be carried out by limiting the field of view of the recording video camera (if appropriate settings are available). Such adjustment, however, requires considerable time and often leads to fatigue of the subject even before the start of the main experiment (Barabanshchikov, Zhegallo, 2014). Also, infrared oculography is more sensitive to changes in the level of illumination in general, which requires constant control of illumination. The main disadvantage of this method is that it can only measure eye movement in the range of about ±35 degrees along the horizontal axis and ±20 degrees along the vertical axis (Fazylzyanova, Balalov, 2014). However, such problems need to be solved not only when registering the pupillary reaction in the infrared, but also in the optical range.

Researchers have found that in a lie situation, changes in pupil size are indeed observed. At present, the direction of deception detection based on the registration of eye movement is actively developing in the United States. Researchers from the University of Utah, USA, used this technique in experiments, the essence of which was to present the subject on the monitor screen with a series of questions that could be answered either truthfully or falsely. By recording the respondent's cognitive responses, which are manifested in changes in pupil diameter, response time, and the number of rereads of the question, the researchers found that cognitive load increases during lying. As a result, prolonged and intense oculomotor activity is observed. For example, when making false statements and deception, there is not only an increase in the time to answer a question, but also an increase in the diameter of the pupil and a decrease in the frequency of blinking. The results of such studies suggest that the registration of oculomotor activity during tests, reading, and answering questions of interest to the employer can be effectively used in the selection of job candidates, since oculomotor activity is difficult to consciously control (Bukhtiyarov et al., 2019).

In the field of technologies for assessing the reliability of information reported by a person, the EyeDetect method, developed in the United States and based on the analysis of eye movements and changes in pupil diameter, should be noted (Vinogradov, Kasperovich, Karavaev, 2018).

In recent years, the Investigative Committee of Russia has been developing a method for revealing hidden information when a person recognizes a relevant stimulus using Eye-Tracker (Zhbankova, Gusev, 2018). According to the results of the research, it was found that the information obtained on the polygraph and Eye-Tracker during personnel examinations coincides by 80.1%, depending on the subject of the hidden information (alcohol abuse, drug use, criminal offenses, and so on) (Selezneva, Voronova, 2021).

Current eye-tracking methods present many challenges, such as identifying fixations, assigning fixations to areas of interest, selecting appropriate metrics, eliminating potential errors in gaze location, and handling scan interrupts. Special considerations are also needed when designing, preparing, and conducting eye-tracking studies (Goldberg, Helfma, 2010). Using pupil detection rather than iris detection will make the system much more robust to people with high iris melanin. The pupil gives a much more precise line of sight than the iris, but is a bit more demanding when it comes to image quality, and also requires flash placement, which minimizes the risk of a bright pupil.

Despite the variety of existing possibilities of the described methods, the pupillary reaction is not used to indicate the psychophysical state in the selection of personnel. And if the selection of workers is online and can apply eye-tracking (using infrared cameras), then this will require additional equipment. This makes it almost impossible to implement eye-tracking remotely. We propose to use a digital video camera not only to register oculomotor activity, but also to register changes in the area of pupil size (obtaining pupillograms). Information about the pupillary reaction in response to stimulus material makes it possible to significantly clarify information about the psychophysical state of a person obtained during ordinary visual contact. A video camera operating in the optical range can remotely and unobtrusively control the dynamics of pupil size, blinking frequency, and oculomotor reaction.

3.3   METHODOLOGY AND TECHNIQUE OF THE EXPERIMENT

The research procedure includes several stages:

- Selection of test objects aimed at identifying a stress/strain state
- Registration of the pupillary reaction when viewing the stimulus material (with the help of a pupillographic module), synchronized with the measurement of the psycho-emotional state by means of the GSR
- Analysis of standard methods and selection of the most suitable ones for obtaining information about the psychological state of the subjects
- Correlation analysis of the results of measurements of the psycho-emotional state of the subjects using standard and pupillographic methods
- Drawing up an algorithm for searching for deviations of a person's psychophysical state from a given norm

If there is a correlation between the results of measuring the psychophysical state of a person by standard methods and the dynamics of the pupillary reaction in response to the presented stimuli, then the identified correlates can be used in the algorithm for searching for deviations in the psychophysical state of a person from a given norm.

3.3.1   SELECTION OF TEST OBJECTS AIMED AT IDENTIFYING A STRESSFUL/ STRAINED STATE AND A CONTINGENT OF EXPERIMENT PARTICIPANTS

Images of various subjects were used as stimulus material. The selection of test objects was carried out based on the fact that there are topics and problems that can cause some kind of response from most people. Test objects were subdivided into two groups: neutral ones that do not carry any information (black dots of different diameters located in different corners of the slide) and informational ones (having an emotional coloring or aimed at drawing attention to any detail of the image). At the same time, the subject of test objects depends on the purpose of the experiment. In this case, identify a stressful/stressful state.

To identify a stressful/tensioned state, it is necessary to use test objects that can cause at least a small amount of psycho-emotional stress that arises as a result of experiencing sufficiently strong emotions (resentment, deceit, danger, threat, information overload, etc.). For operators of various control systems, dispatchers, and other workers of similar professions, there is mainly such a subspecies of psycho-emotional stress as information stress (it consists in an excess of information), which is aggravated by responsibility, making quick and correct decisions.

From the selected images, a survey was compiled in Google forms. Participants were asked to rate the emotions they felt when viewing each image on a 10-point scale, where 1 is the image is very unpleasant and 10 is the image is very pleasant. The study involved students and staff of the university. An analysis of the survey results made it possible to make a presentation of stimuli (images) with the most likely emotional response. The intensity of the most likely emotions evoked by stimuli is low intensity. According to the results of the survey on a scale from 0 to ± 5 (the strongest emotion), slides containing insects evoke an emotional reaction with a probability of $p = 0.42$; the image of tombstones evokes emotion with a probability of $p = 0.64$.

The change in the psychophysical state, due to the impact of the test object, must be considered relative to the state that was before the start of the impact. That is, neutral test objects are needed that do not carry any information, or contain insignificant information for people of a certain category. Such test objects (calibration) make it possible to "adjust" the system to the selected category of persons, allow taking into account spontaneous fluctuations of the measured parameter, random noise.

Several series of experiments were carried out. Thirty people took part in the calibration experiments, males—16 (53.33%), females—14 (46.67%). The average age of the subjects was 20 years. At the same time, subjects with Slavic (60%) and non-Slavic appearance (40%) were present. The control group included nine people, of which six (54%) were males, three (46%) were females. The age of the subjects under 30 years—2 (18%), 30–40 years—2 (18%), over 40 years—5 (45%). At the same time, subjects with Slavic (82%) and Asian appearance (18%) were present. All had normal vision, or corrected glasses. For obvious reasons, there are no official papers, the names and names of employees were not even pronounced.

3.3.2 Research Equipment

All experiments are carried out in a separate room equipped with equipment: a pupillographic module for registering changes in pupil size, an Activationometer 6 hardware and software system, three video cameras, two laptops, a personal computer (PC), and a wall monitor for demonstrating test objects. There are no sources of natural light in the room, a constant temperature and other climatic parameters are maintained. After the participants expressed their voluntary consent to participate in the experiment, they were invited in turn to go to a special room for measurements.

The pupillographic module (Figure 3.1b), designed to record pupillary reaction, consists of the following mandatory components: two ZWOASI-120MC video

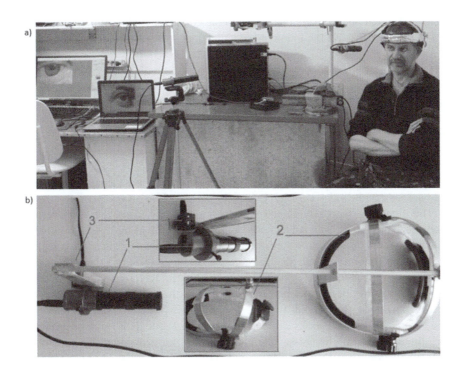

FIGURE 3.1   (a) General view of the measuring equipment. (b) Pupillographic module.

cameras that record changes in pupil size (remote video camera (Figure 3.1a) and a video camera (1) mounted on a thin rod (Figure 3.1b) attached to the helmet (2)). An additional component is a video camera (3) (Figure 3.1b), which records an image on a monitor that displays test objects. If necessary, a video camera is installed that records the readings of the "Activationometer-6" device.

The helmet (2) with fixing straps creates a rigid coordinate connection between the video camera and the head. The rigid frame of the helmet is soldered with a metal plate ring. Two metal plates, curved in a gentle arch, are soldered to it, crosswise. The frame is tightly mounted on a soft helmet and is fixed to it with rivets along the perimeter. In the center of the plates, curved by a gentle arch (at the crosshairs of the frame), a clamp-mount is soldered. In the frontal and occipital parts of the frame, two vertical identical plates with holes are soldered. Several holes are drilled in the plates so that you can choose the height of the video camera by placing the tube in two coaxial holes. A long metal rod is passed through the shackle of the clamp and the corresponding holes in the short plates. A mount for a video camera is soldered to the rod. After selecting the optimal angle of inclination of the video camera (eye in the viewing area), the fixing bolt is tightened. The distance of the video camera from the eye is controlled by shifting the rod along its axis to the desired distance. The rod is moved to the desired distance and tightly clamped in the dome mount. The video camera (1) fixed on the rod is connected to the computer. Video cameras (1) and (2) ZWO ASI 120MC have the following specifications:

- 1X-100X optical zoom lens, focusing range: 20 mm to infinity
- Resolution: 1.2 megapixels 1280 × 960; sensor: 1/3″ CMOS AR0130CS, pixel size: 3.75 μm
- Frame rate: 30 fps at maximum spatial resolution; range exposure: 64 μm– 1000 s
- Baud rate: 12-bit output (12-bit ADC); interface: USB3.0/USB2.0

The video camera (3) (sq8 esp, OV9712 sensor) is located on a small protrusion on the back side of the video camera (1).

The algorithm for studying the psychophysical state using the pupillographic module is as follows. The subject sits comfortably on a specially equipped place, which is a table on which the Activationometer-6 device is located, which registers the galvanic skin response (GSR). A helmet is put on and fixed on the subject. Then the lighting and video shooting parameters are adjusted. Recording on all video cameras and demonstration of video presentations from the monitor screen is turned on. At the same time, the internal time of the cameras is synchronized to thousandths of a second. The subject looks at the monitor screen, while in the process of viewing the stimulus material, he presses the plates of the "Activationometer 6" device and does not release them during the entire video recording process. The contribution to illumination from the monitor from which the stimulus material is shown tends to zero. The luxmeter recorded a change in illumination of ±8 lux, which is below the eye's sensitivity threshold. The experimenter notifies the subject of the completion of the measurement procedure after the video presentation has ended and the black screen turns on. Before the target experiment, the optoelectronic units of the recording system were spatially calibrated. We also conducted experiments to determine the state of the optical system of the eye, with respect to which all subsequent measurements will be considered (calibration state). Calibration frames are inserted into each stimulus material.

To carry out remote studies, we checked the identity of the tracks obtained by a video camera (1), which has a rigid coordinate connection with the head and is located at a distance of 40 cm from the eye plane, and a "free" video camera, located on a tripod at a distance of 250 cm from the eye plane. To check the identity of the results obtained, the subjects had to look steadily at a black dot 6 mm in diameter, while specifically turning their heads. As a result, the attention tracks obtained from both video cameras represented a complex trajectory; therefore, in order to be able to speak with confidence about the direction of a person's gaze, it is necessary to correct the attention tracks. To correct the obtained tracks, we developed a technique based on the transition from the coordinate system associated with a fixed head to the coordinate system associated with the moving center of the pupil (Boronenko, Isaeva, Zelensky, 2021). To do this, the image of the monitor reflected in the cornea was selected on the received video frames, the center of mass of the selection was tracked in each frame and its coordinates were determined, after which both tracks are localized in a small area, corresponding to the observed point. At the 0.01 level, the samples do not differ significantly. Thus, we can assume that remote eye-tracking according to the developed method gives a reliable result.

After passing the study on the pupillographic module, the subjects filled out questionnaires (occupational stress scale, Spielberg–Khanin method, Munsterberg test).

3.3.3 RATIONALE FOR THE CHOICE OF STANDARD TECHNIQUES AND THE PROCEDURE FOR IDENTIFYING AN ALARM CONDITION

There is a hypothesis about the relationship between some characteristics of the working qualities of a person and the main properties of his nervous system (Tikhomirova, 2008). Employees must have a number of qualities that determine their psychological and emotional stability, stress resistance, and the ability to withstand psychophysical stress (Maklakov, Polozhentsev, Rudnev, 1994). To check the conformity of personal qualities, a standardized multifactorial method of personality research (Sobchik, 2002) and a short-term orientation test (Barlas, 2008) are used. For psychodiagnostics of the state of employees, standard methods are used to identify stress/tension and anxiety states: the SAN questionnaire (well-being, activity, mood), the Spielberger–Khanin method, the occupational stress scale, the Luscher color test, the method for assessing neuropsychic stress.

Let us analyze the shortcomings of these methods, which will not allow us to use them to solve the problems of the tasks set. The interpretation of the results of the SAN questionnaire, in order to be objective, should be carried out only by a specialist. When analyzing the state, not only the values of its individual indicators are important, but also their ratio. In a rested person, the assessments of activity, mood, and well-being are usually approximately equal. As fatigue increases, the ratio between them changes due to a decrease in well-being and activity compared to mood (Barlas, 2008). When developing the methodology, the authors proceeded from the fact that the three main components of the functional psycho-emotional state—well-being, activity, and mood—can be characterized by polar assessments, between which there is a continuous sequence of intermediate values. However, there is evidence that the SAN scales are overly generalized (Barkanova, 2009). A significant drawback of the Luscher color test is the lack of mathematical processing, which makes it impossible to compare the results of different researchers and allows arbitrary interpretation (Moskvina, Moskvin, 2013). At the same time, the use of the Luscher color test as an independent technique in the practice of psychological examination, in professional selection, or for solving the problem of personnel assessment is unacceptable (Bleikher, Kruk, Bokov, 1976). In the methodology for assessing neuropsychic stress, a special form of mental state is assessed, which occurs in a person when he is in a difficult (extreme) situation or in anticipation of it (Naiman et al., 2018). The questionnaire includes a list of signs of neuropsychic stress, contains 30 characteristics of this condition. They are divided into three groups according to the severity. Diagnosis is carried out individually, an important condition is good lighting and sound insulation in the room (Wegner D.V., Akimenko G.V., 2021).

When developing the PSM-25 psychological stress scale, the authors eliminated the existing shortcomings of traditional methods for studying stress states, aimed mainly at indirect measurements of psychological stress through stressors or pathological manifestations of anxiety, depression, frustration, etc. The scale describes the state of a person experiencing stress, as a result of which it disappeared the need to identify variables such as stressors or pathologies. The technique is considered universal for application to various age and professional samples in a normal population (Gordeeva, 2010). Numerous studies have also shown that the PSM-25 occupational stress scale has sufficient psychometric properties—correlations of the integral PSM indicator with the Spielberger anxiety scale (r = 0.73) were found (Vodopyanova, 2009). The Spielberger—Khanin anxiety scale allows you to determine the level of both personal and situational anxiety (Gorodetskaya, Konevalova, Zakharevich, 2019). The Münsterberg test is a fairly simple, practical, and easy-to-use test. It is used to diagnose the professional suitability of employees in positions with increased neuropsychic stress. At first glance, the task seems simple. But it successfully fulfills its task—it creates a model of activity, where efficiency depends on the ability to concentrate. The Munsterberg test allows not only to study the selectivity of perceptual attention (Kirdyashkina, 1999), but also to train the concentration of attention.

Based on the analysis of the shortcomings and advantages of existing methods, the following standard methods were selected, used in the selection of personnel and monitoring the current state:

- Scale of psychological stress psm-25 Lemur-Tessier-Fillion
- Spielberger—Khanin anxiety scale (situational and personal anxiety scale)
- Münsterberg test

Also, the choice of these methods is justified by their reliability, ease of use and interpretation of the results, and low time consumption. It is important to note that during the experiment, the subjects had no motive for falsifying answers.

Before the start of the experiments, the participants filled out three questionnaires: the occupational stress scale, the Spielberger–Khanin method (scales of personal and situational anxiety), and the Munsterberg test. The participants filled out the first two questionnaires as they were busy, the time for completing the Munsterberg test was limited—2 minutes. After subsequent measurements, the subjects passed only the Munsterberg test (each time a new one) and filled out the scale of situational anxiety. For obvious reasons, the questionnaires were absolutely anonymous, the sheets indicated only the shift and the time of day of the experiment (day/evening). As a result, a set was to be assembled for each subject, which included a scale of professional stress (1 pc), an anxiety scale—a scale of personal anxiety (1 pc), and a scale of situational anxiety (before and after the shift), the Munsterberg test (before and after shifts).

3.4 PSYCHOPHYSICAL CORRELATES OF STRESS AND AN ALGORITHM FOR ITS DETERMINATION

For remote unobtrusive diagnostics of a stressful state, it is necessary to determine indicators that allow determining its presence. The search for psychophysical correlates was carried out among the parameters obtained by standard methods and pupillographic.

Evaluation of the level of stress in a standard way showed that within the group of subjects, situational anxiety at the beginning and end of the shift does not differ significantly, and attention at the beginning and end of the shift does not differ either. According to the scale of professional stress, all employees have a low level of professional stress, two people had moderate anxiety. For all employees, situational anxiety increased by the end of the shift, but did not go beyond the permissible zone (corresponded to the value of the level of low anxiety). At the same time, the negative correlation with concentration of attention increased. If we turn to individual indicators, the concentration of attention of three employees remained unchanged, for two it decreased, for one employee, on the contrary, the indicators of concentration of attention increased. This can be explained by the fact that, according to physiological characteristics, a person feels better in the evening than in the morning. For two employees, situational anxiety, compared with the beginning of the shift, increased by 8 units. As expected, correlates-indicators of a stressful state, the following are considered:

- *Pupillographic method:* S\Smed—change in pupil size (where Smed—median value of the area of pupil size), ΔI—change in current strength (change in GSR values), n—number of blinks when viewing a specific test object; N is the total number of flashes.
- *Standard method:* a—level of professional stress, b1—level of situational anxiety at the beginning of the shift, b2—level of situational anxiety at the end of the shift, c—level of personal anxiety, d1—level of attention at the beginning of the shift, d2—level of attention at the end of the shift. The results of the correlation analysis are presented in Table 3.1.

The Kendall correlation gives the same results.

The response to stimulus material depends on the significance of the influencing test object. The experiments used stimulus material with different emotional coloring, but of low intensity. Therefore, in a person who was in an emotionally stable state, there was no response to the test object, or it was very weak. Also, the participants in the experiment did not have to solve any tasks that enhance the cognitive activity of the brain. Therefore, the response to cognitive load, and therefore changes in GSR, should be very small. That is, the conditions under which the measurements were taken are close to everyday ones. The closer a person is to a state of stress, the greater changes in psychophysical parameters should be recorded. Especially if the incentives (even very weak ones) are related to the "pain point"—the cause of stress. When viewing a significantly important stimulus, memory is accessed—cognitive processes of comparison and recollection are launched. This leads to an increase in GSR, which is sensitive to cognitive processes (Figure 3.2). If the level of stress is high, then cognitive processes lead to the appearance of emotions. The greater the intensity of emotions, the closer the person to dangerous manifestations, the stronger the increase in S/Smed.

Measurement of the stress level of employees in a mild stress state showed a decrease in attention (d) to an insignificant stimulus material (or task), accompanied

TABLE 3.1
Spearman Correlations

Spearman Corr., p-value	a	b1	b2	c	d1	d2	ΔI	Blinking	S/S	Total Number of Blinks
"a"	1	0,551	0,754	0,058	-0,985*	-0,464	-0,338	0,833*	-0,414	0,581
	–	0,257	0,084	0,913	3,09E-04	0,354	0,512	0,039	0,414	0,228
"b1"	0,551	1	0,662	0,529	-0,529	-0,515	-0,429	0,564	-0,841*	0,132
	0,257	–	0,152	0,28	0,28	0,296	0,396	0,244	0,036	0,803
"b2"	0,754	0,662	1	-0,015	-0,676	-0,662	-0,857*	0,751	-0,735	0,118
	0,084	0,152	–	0,978	0,14	0,152	0,029	0,085	0,096	0,824
"c"	0,058	0,529	-0,015	1	-0,015	-0,221	0,257	0,344	-0,42	-0,368
	0,913	0,28	0,978	–	0,978	0,674	0,623	0,504	0,407	0,473
"d1"	-0,985*	-0,529	-0,676	-0,015	1	0,441	0,257	-0,736	0,315	-0,676
	3,09E-04	0,28	0,14	0,978	–	0,381	0,623	0,095	0,543	0,14
"d2"	-0,464	-0,515	-0,662	-0,221	0,441	1	0,6	-0,344	0,315	0,221
	0,354	0,296	0,152	0,674	0,381	–	0,208	0,504	0,543	0,674
"ΔI"	-0,338	-0,429	-0,857*	0,257	0,257	0,6	1	-0,365	0,612	0,172
	0,512	0,396	0,029	0,623	0,623	0,208	–	0,477	0,196	0,745
"Blinking"	0,833*	0,564	0,751	0,344	-0,736	-0,344	-0,365	1	-0,671	0,188
	0,039	0,244	0,085	0,504	0,095	0,504	0,477	–	0,145	0,722
"S/S"	-0,414	-0,841*	-0,735	-0,42	0,315	0,315	0,612	-0,671	1	0,105
	0,414	0,036	0,096	0,407	0,543	0,543	0,196	0,145	–	0,843
"Total Number of Blink"	0,58	0,132	0,118	-0,368	-0,676	0,221	0,172	0,188	0,105	1
	0,228	0,803	0,824	0,473	0,14	0,674	0,745	0,722	0,843	–

2-tailed test of significance is used
*: Correlation is significant at the 0.05 level

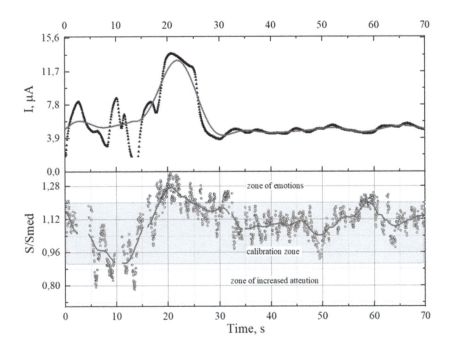

FIGURE 3.2 Synchronized measurements of GSR-$I(t)$ and pupillogram- $S(t)$ taking into account the division into zones: calibration zone "0"; attention zone "1"; emotional zone "2".

by a slight increase in anxiety (a, c, b). Fatigue leads to a similar result, which is confirmed by the presence of a negative correlation. The representation of the measured parameters in binary code is presented in Table 3.2.

Researchers have proven (Vrana, Spence, Lang, 1988) that there is a negative correlation between the sign of emotion and the frequency of blinking, that is, the more unpleasant the stimulus, the stronger the blinking. It was also proved (Bradley, Cuthbert, Lang, 1990) that there is a positive correlation between the sign of emotion and the latent period of blinking, that is, unpleasant stimuli cause faster blinking reactions. However, it is important to take into account the strength of emotions, since the effect of the sign of emotion on blinking is observed only when the emotion has a high strength (Kosonogov, Martinez-Selva, Sanchez-Navarro, 2017).

In the area of low stress or its absence, GSR and S/Smed should go in antiphase with stress indicators and be consistent with the work of the cognitive area, because the brain works well, nothing interferes with concentration (GSR and S/Smed—positive correlation). The stronger the stress, the weaker the emotional stability, the less the brain works, GSR and S/Smed in antiphase. The algorithm for determining the stress state is shown in Figure 3.3.

The algorithm for determining the stress state includes two parallel processes that run simultaneously. The first process is a demonstration of test objects on the monitor screen, the second one searches for respondents—people who have begun to look at the monitor. The respondent search cycle continues until the moment of

TABLE 3.2

Representation of Measured Parameters in Binary Code

Measured parameter track in area of interest (binary code)	no stress			Stress is small			"Pain point"		
	Stimulus significance			Stimulus significance			Stimulus significance		
	target	not target	reaction time	target	not target	reaction time	target	not target	duration, special reaction
	0	1	0	0	1	1	0	1	1
ΔI	0	1		0	1	0	0	1	
					↑			↑	
S/S (zone)	0	1	-	1	0	-	0	2	+
					↑		↑	↑	
S/S (binary code)	0	0	0	0	1	0	1	1	1
n	↑		↓	↑	↓	↑		↓	↑
n (binary code)	0	1	1	0	1	0	1	1	1

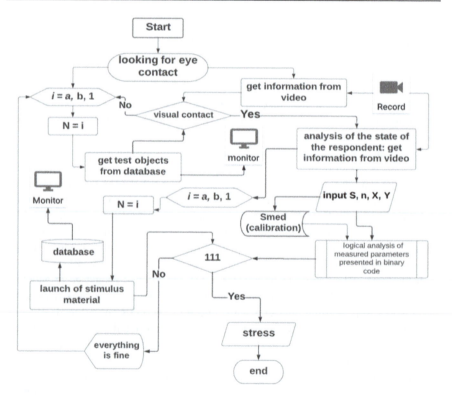

FIGURE 3.3   The algorithm of the system for remote monitoring of the psychophysical state of employees.

discovery. After visual contact is detected, the process of obtaining information from the video (reading the pupil), determining the attention track and the number of blinks is started. This starts a new cycle of demonstration of stimulus material. The theme of the stimuli corresponds to the theme of the test object that aroused the respondent's interest (visual contact). All measurements are individually normalized to the median (calibration) value of the pupil size. Each of the measured parameters, based on the value, according to Table 3.2, is presented in binary form. If all measured parameters are "1", then the stress state is detected. When a stressful state is detected, a final process occurs, such as notifying the experimenter; if the state is not detected, then the cycle returns to the beginning to search for a new respondent.

3.5   RESULTS AND DISCUSSION

The results of the correlation analysis confirmed the presence of a correlation between the results of standard methods and the proposed one. There is a negative correlation of situational anxiety at the beginning (−0,63) and at the end of the shift (−0,74) with a change in pupil size. Correlation of the same parameters with changes in GSR (−0,51) and (−0,86), respectively. According to the results of standard methods, there were no violations in the psycho-emotional state of employees (there is no professional stress, anxiety and concentration of attention are within the normal range), which is also confirmed by assigning the obtained pupillograms to the zero zone (calibration zone). At low levels of stress, cognitive abilities remain normal (a person is not nervous and does not experience a feeling of anxiety), there is an increase in GSR values; however, the resulting pupillograms will be in the first zone (zone of attention), since the changes are not caused by an emotional response, but by examining some then the image details.

The mean value S/Smed of the emotional reaction differs from the mean values of the calibration and recognition reactions by more than 3σ (Figure 3.4). This means that the signal is reliably recognized. The absence of a clear boundary between the states in the pupillogram is explained by the presence of a transient process. The manifestations of the components of the mental state are individual. Under the influence of stimuli (circumstances), after a certain latent period, the components of the psychological state begin to change. If the activity is above or below the average level, then the state is defined as nonequilibrium. The instability of the state is estimated by the rate of its destruction and the depth of changes.

If viewing the test object leads to a change in the size of the pupil in connection with the experienced emotions, then the pupillograms will belong to the second zone (emotional zone) (Isaeva, Boronenko, Boronenko Yu, 2021; Boronenko et al., 2021). Therefore, the size of the pupil can become a criterion for remote diagnostics/monitoring of the psychophysical state of employees in the course of their professional activities. Using eye-tracking synchronized with the pupillary reaction, you can get a more complete analysis of the current state of the employee. In this case, the value of such a parameter as the frequency of blinking will also be important. There are several types of blinking: arbitrary, reflex, and periodic. Voluntary and reflex closure of the eyelids are not of interest in terms of study. Periodic blinking, on the contrary, is of considerable interest to psychophysiologists. In time, the flashing lasts 0.35 s.

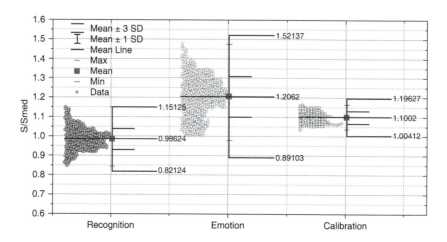

FIGURE 3.4 Significant difference in pupillary response: emotion, attention, calibration (confirms Friedman Analysis at 0.01).

The frequency of periodic flashing in different people can be very different, but for each individual it is quite stable under constant conditions. In different people, the frequency of blinking varies from 1 to 46 per minute. Although the neurophysiology of periodic blinking is still poorly understood, it is safe to say that the frequency of blinking varies depending on the state of the psyche and reflects the "level of mental stress" (Hall, Cusack, 1972).

Researchers from Sweden conducted experiments to assess the actions of police officers in stressful situations (Bertilsson, 2020). They found that the average pupil size when performing test tasks (the scenario of which is realistic) with a moderate threat did not significantly differ from the diameter of the pupil recorded when performing a task with a high level of threat. However, the fact that the authors did not find a difference between the reaction of the pupil to weak and strong threats can be explained as follows. First, the participants knew that they were participants in the experiment, so all tasks were most likely processed as cognitive, which is confirmed by their second finding (correlation of pupil size with task performance). And, most likely, all participants did not have emotional burnout, i.e., for them, the level of stress was acceptable (Bertilsson et al., 2019). When implementing the proposed method in real conditions, the difficulty will be in the selection of additional illumination parameters. It is necessary to create conditions under which the result will not depend on factors such as lighting, background, camera angle, and video camera resolution. If it is impossible to exclude additional illumination from the monitor on which test objects are displayed, it is necessary to use equal-brightness stimuli.

3.6 CONCLUSION

An analysis of the methods used in hiring to check the suitability of applicants for a position in the Ministry of Emergency Situations showed the following:

- The applied methods allow controlling emotional stability, analytical abilities, overcontrol, pessimism, emotional lability, impulsivity, femininity/masculinity, rigidity, anxiety, individuality, optimism, introversion, and analytical abilities; however, they have a number of disadvantages: the subjectivity of information and the possibility of planned answers.

Analysis of standard methods for monitoring the condition of employees of the Ministry of Emergency Situations showed the following:

- The methods used allow you to control the current psycho-emotional state: stressful, tense, anxious (SAN method, Spielberger–Khanin method, and occupational stress scale); however, test results can be distorted by pre-planned answers. To prevent the professional health of employees, psychodiagnostic monitoring is carried out every two years (Vasilchenko, Turova, Stabrovskaya, 2020). However, the presence of suicides indicates the imperfection of the measures taken. However, it is impossible to constantly monitor psychological stability using standard methods.

In connection with the need to improve the means of monitoring the psychological state, a technique has been developed that allows monitoring the state of a person who is in the field of view of video cameras.

Approbation of the developed methodology was carried out on employees of the Ministry of Emergency Situations. Correlation analysis of research results using standard methods and the proposed one showed significantly significant ($p = 0.05$) correlations:

- Inverse correlation (−0,93) d1 (attention concentration at the beginning of the shift) and b2 (indicators of situational anxiety at the end of the shift)—poor attentiveness at the beginning of the shift leads to the accumulation of anxiety by the end of the shift.
- Correlation dependence (0,85) ΔI (change in GSR values) and d1 (attention concentration at the beginning of the shift)—with stimulus material that does not carry a cognitive load and is aimed at obtaining a weak emotional response. A positive correlation between GSR and attention, with a decrease in the number of blinks, indicates a person's cognitive reaction outside the "pain zone".
- Positive correlation between the level of occupational stress and the total number of blinks (0,83). The more the stress, the greater the total number of blinks, since it is chronic stress and increased levels of anxiety that are the primary causes of ocular neurosis.
- An algorithm for the operation of the system for remote monitoring of the psychophysical state of employees is proposed.

The proposed method can become the basis for unobtrusive noncontact control of a person's psychophysical state.

REFERENCES

Barabanshchikov V. A., Zhegallo A. V. (2014). *Eyetracking: Methods for recording eye movements in psychological research and practice.* Moscow: Kogito-Centre. p. 128.

Barkanova O. V. (2009). *Methods for diagnosing the emotional sphere: Psychological workshop.* Krasnoyarsk: Litera-Print. pp. 205–210.

Barlas T. V. (2008). The use of a short selection test in solving practical problems of education. *Bulletin of the Moscow State Linguistic University: Education and Pedagogical Sciences.* 539. pp. 47–56.

Bertilsson J. et al. (2019). Stress levels escalate when repeatedly performing tasks involving threats. *Frontiers in Psychology.* 10(1562).

Bertilsson J. et al. (2020). Towards systematic and objective evaluation of police officer performance in stressful situations. *Police Practice and Research.* 21(6). pp. 655–669.

Bleikher V. M., Kruk I. V., Bokov S. N. (1976). Clinical pathopsychology. *Medicine of the Uzbek SSR.* pp. 57–142.

Boronenko M. P., Isaeva O. L., Boronenko Y., Zelensky V., Gulyaev P. (2021). Recognition of changes in the psychoemotional state of a person by the video image of the pupils. In: Wojtkiewicz K., Treur J., Pimenidis E., Maleszka M. (eds) *Advances in computational collective intelligence. ICCCI 2021: Communications in computer and information science*, vol. 1463. Cham: Springer. https://doi.org/10.1007/978-3-030-88113-9_10.

Boronenko M. P., Isaeva O. L., Zelensky V. I. (2021). Method for increasing the accuracy of tracking the center of attention of the gaze. *2021 International Symposium on Electrical, Electronics and Information Engineering.* pp. 415–420.

Bradley M. M., Cuthbert B. N., Lang P. J. (1990). Startle reflex: Emotion or attention? *Psychophysiology.* 27(5). pp. 513–522.

Bukhtiyarov I. V. et al. (2019). New psychophysiological approaches used in the professional selection of candidates for dangerous professions. *Occupational Medicine and Industrial Ecology.* 59(3). pp. 132–141.

Fazylzyanova G. I., Balalov V. V. (2014). Eye-tracking: Cognitive technologies in visual culture. *Bulletin of Russian Universities: Mathematics.* 19(2). pp. 628–633.

Goldberg J. H., Helfman J. I. (2010). Comparing information graphics: A critical look at eye tracking. *Proceedings of the 3rd BELIV'10 Workshop: Beyond Time and Errors: Novel Evaluation Methods for Information Visualization.* pp. 71–78.

Gordeeva T. O. (2010). Development of the Russian version of the dispositional optimism test (LOT). *Psychological Diagnostics.* 2. pp. 36–64.

Gorodetskaya I. V., Konevalova N. Y., Zakharevich V. G. (2019). Study of situational and personal anxiety of students. *Bulletin of the Vitebsk State Medical University.* 5(18). pp. 120–127.

Hall R. J., Cusack B. L. (1972). The measurement of eye behavior: Critical and selected reviews of voluntary eye movement and blinking. *U.S. Army Technical Memorandum* No. 18–72. Maryland: Aberdeen Proving Ground, Human Engineering Laboratory, Aberdeen Research and Development Center.

Isaeva O. L., Boronenko M. P., Boronenko Y. V. (2021). Making decisions in intelligent video surveillance systems based on modeling the pupillary response of a person. *2021 IEEE 6th International Conference on Computer and Communication Systems (ICCCS).* pp. 806–811. https://doi.org/10.1109/ICCCS52626.2021.9449315.

Kirdyashkina T. A. (1999). Methods for the study of attention. *Practicum in Psychology Textbook Chelyabinsk Izd.* 72.

Kosonogov V. V., Martinez-Selva J. M., Sanchez-Navarro J. P. (2017). Review of modern methods for measuring physiological signs of the sign and strength of emotional states. *Theoretical and Experimental Psychology.* 3.

Kotlyachkov O. V. et al. (2017). Improving the recruitment system. *Fotinskiye Readings.* 2. pp. 90–95.

Kuznetsova L. E. (2017). Theoretical analysis of the problem of psychological support for employees of the Ministry of the Russian Federation for Civil Defense, emergencies and disaster relief with post-traumatic stress disorder. *Modern Psychology: Materials of the V Internal Scientific Conference (Kazan, October 2017)*. pp. 31–37.

Maklakov A. G., Polozhentsev D. A., Rudnev D. A. (1994). Psychological mechanisms of type A behavior in young people during the period of adaptation to long-term psycho-emotional stress. *Psikhologicheskie Zhurnal*. 14(6). pp. 86–94.

Moskvina N. V., Moskvin V. A. (2013). Diagnostics of regulation processes in sports psychology. *Ministry of Sports of the Russian Federation Department of Education of the City of Moscow*. p. 336.

Naiman A. B. et al. (2018). The influence of neuropsychic stress on the emotional status and psychophysiological characteristics of search and rescue service employees: Master's thesis in the field of study. *Psychology of Safety and Health*. 7(4). pp. 137–150.

Selezneva E. I., Voronova A. A. (2021). Eye-tracking as an alternative or additional lie detection technology to the polygraph. *Perspectives: collection of scientific articles undergraduates and graduate students*. 10. Scientific editor Z.Kh. Saralieva., N.Novgorod, pp. 154–159.

Shlenkov A. V. (2005). Psychological control and correction of negative mental states among employees of the state fire service of the ministry of emergency situations of Russia. *Bulletin of Psychotherapy*. 13. pp. 76–87.

Sobchik L. N. (2002). *Standardized multifactorial method for studying personality SMIL*. St. Petersburg: Speech.

Tikhomirova O. V. (2008). Individual-psychological features of the personality in the professional activity of employees of the ministry of emergency situations (problem setting. *Society and Law*. 1(19). pp. 286–289.

Vasilchenko N. V., Turova N. N., Stabrovskaya E. I. (2020). Study of the influence of individual psychological peculiarities on the safe behavior of employees of EMERCOM of Russia. *Scientific and Analytical Journal "Bulletin of St. Petersburg University of the State Fire Service of the Ministry of Emergency Situations of Russia"*. 4.

Vinogradov M. V., Kasperovich Y. G., Karavaev A. F. (2018). Improving psychophysiological methods for assessing the reliability of information reported by a person. *Psychopedagogy in Law Enforcement Agencies*. 4(75). pp. 96–102.

Vodopyanova N. E. (2009). *Psychodiagnostics of stress*. St. Petersburg: Peter. p. 336.

Vrana S. R., Spence E. L., Lang P. J. (1988). The startle probe response: A new measure of emotion? *Journal of Abnormal Psychology*. 97(4). pp. 487–491.

Wegner D. V., Akimenko G. V. (2021). Stress resistance as a factor in the development of a positive attitude towards learning activities among students. *Diary of science*. 1 (49). Chief editor: Mukhin M.N., Perm., pp. 17–28.

Yuzhakov M. M., Avdeeva D. K., Nguyen D. K. (2015). A review of methods and systems for studying human emotional stress. *Modern Problems of Science and Education*. 2(2). p. 134.

Zhbankova O. V., Gusev V. B. (2018). Application of eye-tracking in the practice of professional selection of personnel. *Experimental Psychology*. 11(1). pp. 156–165.

4 Ergonomic Work from Home Recommendations Using TRIZ

"Stop Robbing Peter to Pay Paul"

Poh Kiat Ng*, Peng Lean Chong, Jian Ai Yeow, Yu Jin Ng, and Robert Jeyakumar Nathan
* Corresponding author

CONTENTS

DOI: 10.1201/9781003383444-4

4.1 INTRODUCTION

From the equivocal working conditions sparked by COVID-19, a new norm has surfaced in the work environment, namely, the work from home (WFH) concept. Even before the COVID-19 pandemic, the WFH concept has been explored as a means of improving work–life balance [1]. However, the notion that WFH practices are able to considerably alleviate work stress may not be entirely fool-proof.

According to a survey by the McKinsey Global Institute, several companies are ready to shift to flexible workspaces after the global pandemic [2]. The survey, conducted among 278 executives, found that on average, respondents were looking to reduce office space by 30%. However, the survey results indicate that work areas with a high level of physical interaction are likely to change more post-pandemic, with a forced shift to online or other digital transactions, which may lead to a change in behavior in daily work. While there are many factors involved, some work-oriented factors worth exploring include challenges related to employer's perception, work schedule arrangements, improper work environment, and the lack of face-to-face communication.

WFH studies mostly fall under the domain of human resources management (HRM) [3–5]. Under the Malaysian Occupational Safety and Health Act 1994 [6], employers are required to ensure the safety, health, and welfare of all their employees in the workplace as far as is reasonably practicable. The term "Place of Work" is defined under OSHA 1994 as the "premises where persons work", and "premises" include any land, building or part of any building, and any tent or movable structure. The Malaysian Minister of Human Resources had made an amendment to the Employment Act 1955 on flexible work or WFH policy. The scope of the term "premises" includes "work-related activities under its control and in the nature of work" and the employer is only liable for WFH violations if its employees are engaged in something related to their work at the time of the violation. The Chief Executive Officer of the Social Security Organisation (SOCSO) has extended accident insurance to WFH situations, although proving the cause of the injury is still a gray area.

Although there are HRM policies that support WFH practices, there are few studies that examine the process that leads to the proposal of such policies. While there are studies that have investigated the use of the theory of inventive problem-solving (TRIZ) as a creative method to solve HRM problems [7–8], its use in solving WFH problems remains unknown. This study aims to propose ergonomic recommendations for potentially addressing WFH challenges using TRIZ.

4.1.1 "THEFT-PROOFING PETER" WITH TRIZ

The contradictions between WFH practices and work–life balance can be explored and possibly resolved with TRIZ tools and techniques. TRIZ is a Russian acronym for the theory of inventive problem-solving. It is a problem-solving philosophy based on logic, data, and research, rather than on intuition [9]. It uses the knowledge and ingenuity of thousands of engineers to accelerate creative problem-solving for project teams. The approach brings repeatability, predictability, and reliability to the problem-solving process and provides a set of reliable tools [10].

The work–life balance research problem can be analogized with the expression "robbing Peter to pay Paul", which means to take from one merely to give to another. In his paper, Dempsey [11] used this analogy to describe the following contradiction between production engineering and ergonomics:

> From a production engineering standpoint, production bottlenecks can be alleviated by increasing throughput at that point in the production chain which may be undesirable from an ergonomics standpoint. When ergonomics and production engineering function separately, there is always the risk that gains in productivity provided by the ergonomists will be robbed by the production engineers resulting in an increase in pace, ultimately resulting in a net increase in the repetitiveness of a job. Repetitiveness can contribute to the risk of work-related musculoskeletal disorders.
>
> [11, p. 172]

Using the analogy by Dempsey [11], one could assume that the employers are production engineers (or "Paul"), and the WFH practitioners (employees) are the ergonomists (or "Peter"). Employers may see WFH practices as an opportunity to raise targets or expectations because of a misconception that employees have more than enough time at home to achieve them. Such a misconception is also based on the assumption that every home is a retreat from the hustle and bustle of the office.

This research aims to propose "theft-proof" recommendations for "Peter" with TRIZ while still being able to pay "Paul". In principle, with TRIZ algorithms, it might be possible to retain the positive aspects of WFH practices while reaping the benefits of work–life balance.

4.1.2 WORK FROM HOME

The concept of working from home (WFH) is a work practice where workers can benefit from not having to commute to an office or a central workplace in some companies, but can focus on doing their daily work at home to achieve a work–life balance. This practice became the norm in many companies with the outbreak of the COVID 19 pandemic, as social distancing was used to control the COVID-19 infection rate [12].

As the world gradually becomes accustomed to the implementation of WFH, it becomes increasingly important to find the right balance between managing work and family commitments, as such a balance directly or indirectly affects an employee's job satisfaction, time with family, organizational commitment, and physical and mental well-being [13]. Employees who have a good work–life balance significantly improve psychological well-being at work and the work environment by preventing burnout, contributing better to goal achievement, and maintaining a positive attitude toward work [14]. Gorgenyi-Hegyes, Nathan [15] suggested that this finding was also evident during the COVID-19 pandemic where organizations that promoted employee well-being were found to significantly improve their employees' physical, mental, and emotional health, which eventually led to enhanced employee

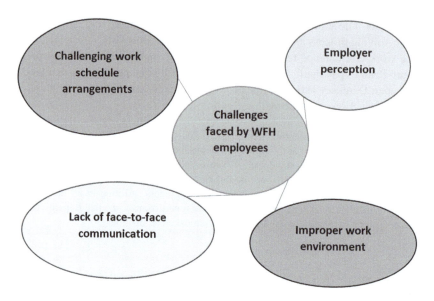

FIGURE 4.1 Possible challenges with WFH.

satisfaction. According to an article by Apollo Technical [16], Stanford University carried out a study involving 16,000 workers over the past nine months, and concluded that working from home increased productivity by 13%.

Despite the aforementioned advantages, there has yet to be a direct link established between employees' well-being and their loyalty to their workplace during the COVID-19 pandemic. Creating ideal work–life balance conditions with WFH is not an easy task, as workers often face numerous challenges when working at home. Some of the possible challenges are shown in Figure 4.1.

4.1.2.1 Employer Perception

A critical dilemma faced by employees on WFH includes the expectation by employers to work longer hours [17–18]. A local newspaper in Malaysia reported that many workers had to put up with harsh treatment from demanding, insensitive employers during the WFH period [19]. It was also reported that some workers received angry, emotional emails or chat messages that made the situation worse. Workers practicing WFH might be asked by their employer to opt for longer working days to compensate for their lack of physical interaction, easy availability at work, and presence in the office. These employees may feel pressured to agree to such terms, fearing that their refusal may be seen as an excuse to avoid work in order to spend more personal time at home. As the line between work and non-work becomes blurred [20], these employees may work well beyond their regular hours and have a heavier than normal workload.

Micromanagement also poses challenges to employees when supervisors push their staff to constantly report on work, which leads to work fatigue [21]. Some employers may micromanage their employees because of a lack of confidence in their abilities to make continuous work progress. Without the possibility of face-to-face meetings, employees practicing WFH may have no choice but to comply with requirements to hold longer and more frequent virtual meetings than usual to enable micromanagement activities. Such employees would find themselves moving from one virtual meeting to another with a major portion of the day spent on attending meetings. This predicament may be especially true for those who need to report to multiple job supervisors. virtual meetings should be limited to a reasonable time frame, preferably no longer than 30 minutes. In addition, employers should allow the video camera to be turned off on certain occasions. If these guidelines are not followed, some employers may equate the act of "turning off the camera" for a virtual meeting with "being absent" or "not paying attention".

4.1.2.2 Challenging Work Schedule Arrangements

With family and household matters coming in the way of work-related matters within the standard office hours, work schedule arrangements for employees at all levels become increasingly erratic, resulting in the extension of work beyond regular office hours. Such a situation can jeopardize both employee motivation and family life.

From an employee's perspective, the most crucial challenge faced with WFH includes blurred boundaries between work and personal life [22]. It seems that almost 70% of professionals who have switched to remote work because of the pandemic say they now work weekends, and 45% say they regularly work more hours during the week than before [23]. For many, home is a safe and personal place to relax or a sanctuary away from the hustle and bustle of work life. It is a place where the mind calms and the body rejuvenates after a long day at work. With WFH, work commitments have begun to disrupt the boundaries of private life. This creates a culture shock for employees, who become increasingly stressed by the blurring boundaries between work and private life. With the removal of the time clock in the office, some staff are expected to support work beyond the normal working hours of 9 a.m. to 5 p.m. Staff are also expected to be almost constantly connected to their work, reporting to their colleagues or supervisor on their work progress most days and even on weekends for ad hoc work requests.

4.1.2.3 Improper Work Environment

Another dilemma faced by WFH employees is the distractions of working at home [24–26] that prevent them from doing their jobs effectively. Even if employees have a dedicated workspace and work schedule set up for WFH, it can be challenging to avoid distractions while working at home. Some employees have to deal with interruptions and unexpected breaks to attend to urgent domestic duties such as housework, cooking, and childcare while doing their work in a WFH environment. According to Maurer [23], working parents are more likely to work weekends and more than 8 hours a day than working parents without children. Men were more likely than women to report working weekends and having a 40-plus hour week. This challenge negatively impacts the employee's ability to focus on work and decreases work productivity over time. Moreover, family members at home do not hesitate to interrupt them in personal matters by assigning or reminding them of chores or tasks for the day.

Inadequate equipment for an effective work environment [27–28] may also contribute to the challenges faced by workers practicing WFH. Some workers require certain hardware or software in the workplace to perform their work properly. These needs are even more critical for those who deal with specialized technical equipment, machinery, laboratory apparatus, or controlled software to perform their work tasks. The lack of proper equipment also prevents workers from solving organizational problems remotely, such as manufacturing and production problems in factories.

4.1.2.4 Lack of Face-to-Face Communication

The WFH practice also introduces challenges in work communication and coordination among employees and management [29–30]. Kramer and Kramer [31] mentioned that among the most important aspects missing in WFH culture are informal face-to-face meetings, the joy of traveling, and breaking the routine of staying in one place. Based on Hofstede's cultural dimensions theory, Asia is a country with high collectivism where it might be challenging for most citizens to practice WFH without face-to-face meetings or travel [32]. In a normal office environment, it can be difficult enough to hold a meeting to coordinate the different characteristics of team members and align tasks toward a common goal. The challenge becomes even greater at WFH when it comes to ensuring that everyone in a virtual space is communicating effectively on the same page. People also rely on nonverbal communication to connect with each other, and the use of emails, phone calls, and virtual video conferencing alone may not be enough to avoid misinterpretation of work instructions or create a mutual understanding of a common goal [33]. Misinterpretation of messages leads to further challenges in creating a conducive work environment. In short, the lack of face-to-face communication does not promote more "human" innovative problem-solving approaches.

4.1.3 CONSEQUENCES OF WFH CHALLENGES

All of the aforementioned factors have the potential to affect the overall mental well-being of employees [34]. The challenges of WFH can exacerbate the existing work environment, which is already fraught with stress and hurdles, to the point where employees' mental health is severely affected.

4.1.3.1 Reduced Personal or Family Time

Without a work–life balance and with an immense workload at the office and at home, employees practicing WFH may spend most of their personal or family time working through their workload instead of resting and recovering from stress. According to Lebow [35], COVID-19 and the associated procedures of shelter-in-place have had an influential impact on all families but have added special meanings in the context of families contemplating divorce, divorcing, or carrying out post-divorce arrangements. Furthermore, the feeling of being on duty almost all the time makes it challenging for employees in segregating their time for personal rest and family. Employees who frequently allow their work to invade into their personal and family time subsequently succumb to consequences such as burnout, poor personal well-being, and even a broken family.

4.1.3.2   Fatigue and Distrust

The long hours spent by staff practicing WFH in excessive meetings can be tiring and draining for them, leading to a counterproductive effect of lower work efficiency as most of the time is spent attending meetings rather than performing the required tasks. The lack of face-to-face communication between staff in an organization practicing WFH also reduces networking and interaction between them, resulting in greater social barriers, lower team spirit, lower trust, lower motivation, and lower morale among staff [27, 33, 36]. The idea that virtual meetings should help connect employees seems counterproductive to employees' social needs. Employees who are used to interacting and working in a team in real life can lose that sense of shared humanity and trust when things go virtual. In office environments, employees build trust through informal interactions. When using a virtual system, there is a risk that trust between colleagues is lost because employees no longer work together.

4.1.3.3   Distractive and Unsuitable Work Environment

Distractions from family, as well as employees' inability to balance work and personal tasks at home, cause unnecessary stress and frustration as they may not be able to meet their deadlines and fulfill their potential at work. In addition, if certain equipment is not available in the WFH environment, workers face major challenges in overcoming technical limitations that significantly delay their work. The lack of a proper workstation or essential office infrastructure, such as a desk or a suitable sitting workstation at home, could be uncomfortable for staff who have to sit and work for long hours, leading to the risk of musculoskeletal disorders [37–39].

4.2   METHODOLOGY

The factors that point to the challenges faced by workers practicing WFH require critical and immediate attention. Therefore, this chapter proposes to use the theory of inventive problem-solving (TRIZ) to resolve the contradictions in work–life balance for workers practicing WFH.

TRIZ is an approach used in identifying the contradictions and conflicts of a system to resolve problems that are inventive by nature [9, 40]. An inventive problem is a problem that includes a contradiction or a problem for which the path to a solution is unknown [41].

In this study, the contradictions will be formulated based on the WFH challenges identified from the literature review using the "If-Then-But" statement which comprises a manipulative variable change statement (If), a statement with a positive responding variable (Then), and a statement with a negative responding variable (But) [10, 42]. The positive and negative responding variable statements will then be linked with some of the 39 system parameters of TRIZ.

The positive and negative (or improving and worsening) parameters will then be intersected in a matrix known as the contradiction matrix, which is organized in the form of 39 improving parameters and 39 worsening parameters (a 39 × 39 matrix), with each cell entry comprising the most frequently used inventive principles to resolve the formulated contradiction [43].

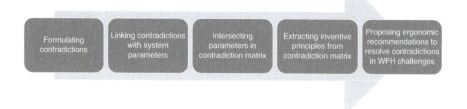

FIGURE 4.2 Process flow of TRIZ methodology.

The TRIZ inventive principles are a list of known solutions extracted from past and existing solutions to inspire inventors and researchers in solving new problems and imagining innovative solutions [44]. The principles extracted from the contradiction matrix will be used as guides to propose feasible policies or solutions to contradictions identified in WFH challenges. Figure 4.2 shows the process flow of the TRIZ methodology in this study.

4.3   RESULTS AND DISCUSSION

4.3.1   FORMULATING CONTRADICTIONS

Based on the literature review, the following factors that challenge employees in practicing WFH were identified.

- Employer perception
- Challenging work schedule arrangements
- Improper work environment
- Lack of face-to-face communication

Using the "If-Then-But" statement as a method to formulate the contradiction (C), the following contradiction statements were proposed.

C1: If employers expect employees who practice WFH to be constantly more engaged in work than usual, then the level of activeness in the employees' workflow increases, but employees will become highly stressed.

C2: If employees who practice WFH are expected to work longer hours than usual, then the amount of accumulated work might decrease over time, but employees will lose out in their personal and family time.

C3: If the working environment for employees who practice WFH does not match the usual working environment, then employees can work at the comfort of their homes, but they will not be able to fulfill all their work demands due to distractions and lack of equipment.

C4: If the employees have lesser face-to-face interactions, then they might avoid possible physical harassment or intimidation at the workplace, but the lack of trust among employees will increase.

4.3.2 LINKING CONTRADICTIONS WITH SYSTEM PARAMETERS

The closest improving and worsening parameters out of the 39 system parameters were identified for every contradiction statement. The positive responding variable (Then) statement was linked with the improving parameter, while the negative responding variable (But) statement was linked with the worsening parameter. Table 4.1 summarizes the list of system parameters linked with the negative and positive responding variables of the contradiction statements.

TABLE 4.1
List of System Parameters Linked with Negative and Positive Responding Variables of Contradiction Statements

WFH Challenge	Contradictions	Positive (+) and Negative (−) Responding Variables	Improving and Worsening System Parameters
Employer perception	C1: If employers expect employees who practice WFH to be constantly more engaged in work than usual, then the level of activeness in the employees' workflow increases, but employees will become highly stressed	(+) then the level of activeness in the employees' workflow increases	Parameter 19: Use of energy by moving object
		(−) but employees will become highly stressed	Parameter 11: Stress or pressure
Challenging work schedule arrangements	C2: If employees who practice WFH are expected to work longer hours than usual, then the amount of accumulated work might decrease over time, but employees will lose out in personal and family time	(+) then the amount of accumulated work might decrease over time	Parameter 26: Quantity of substance
		(−) but employees will lose out in personal and family time	Parameter 25: Loss of time
Improper work environment	C3: If the working environment for employees who practice WFH does not match the usual working environment, then employees can work at the comfort of their homes, but they will not be able to fulfill all their work demands due to distractions and lack of equipment	(+) then employees can work at the comfort of their homes	Parameter 33: Ease of operation
		(−) but they will not be able to fulfil all their work demands due to distractions and lack of equipment	Parameter 39: Productivity
Lack of face-to-face communication	C4: If the employees have lesser face-to-face interactions, then they might avoid possible physical harassment or intimidation at the workplace, but the lack of trust among employees will increase	(+) then they might avoid possible physical harassment or intimidation at the workplace	Parameter 30: Object-affected harmful factors
		(−) but the lack of trust among employees will increase	Parameter 27: Reliability

4.3.3 Intersecting Parameters in Contradiction Matrix

The improving and worsening parameters were intersected in the contradiction matrix to obtain the solution models comprising the inventive principles. Table 4.2 shows the list of inventive principles extracted from the parameters intersected in the contradiction matrix. With reference to Yeoh [10] and Yeoh, Yeoh [42], the definitions of each extracted inventive principle can be seen in Table 4.3.

4.3.4 Proposing Ergonomic Recommendations to Resolve Contradictions in WFH Challenges

Most of the definitions for the extracted inventive principles are technical by nature. Therefore, it was only possible to make use of the relative meanings of the principles in regard to management-related practices or examples. For this study, some of the management-related examples were referenced from Yeoh [10]. Due to the technical

---

TABLE 4.2
List of Inventive Principles Extracted from the Intersection of System Parameters within the Contradiction Matrix

WFH Challenge	Positive (+) and Negative (−) Responding Variables	Improving and Worsening System Parameters	Inventive Principles
Employer perception	(+) then the level of activeness in the employees' workflow increases	Parameter 19: Use of energy by moving object	#23: Feedback #14: Curvature #25: Self-service
	(−) but employees will become highly stressed	Parameter 11: Stress or pressure	
Challenging work schedule arrangements	(+) then the amount of accumulated work might decrease over time	Parameter 26: Quantity of substance	#35: Parameter changes #38: Strong oxidants #18: Mechanical vibration
	(−) but employees will lose out in personal and family time	Parameter 25: Loss of time	#16: Partial or excessive actions
Improper work environment	(+) then employees can work at the comfort of their homes	Parameter 33: Ease of operation	#15: Dynamics #1: Segmentation
	(−) but they will not be able to fulfill all their work demands due to distractions and lack of equipment	Parameter 39: Productivity	#28: Mechanics substitution
Lack of face-to-face communication	(+) then they might avoid possible physical harassment or intimidation at the workplace	Parameter 30: Object-affected harmful factors	#27: Cheap short-living objects #24: Intermediary #2: Taking out
	(−) but the lack of trust among employees will increase	Parameter 27: Reliability	#40: Composite materials

TABLE 4.3

Definitions of Each Extracted Inventive Principle

Principle	Definitions
#23: Feedback	1. Introduce feedback (referring back, cross-checking) to improve a process or action 2. If feedback is already used, change its magnitude or influence
#14: Curvature	1. Instead of using rectilinear parts, surfaces, or forms, use curvilinear ones; move from flat surfaces to spherical ones; from parts shaped as a cube (parallelepiped) to ball-shaped structures 2. Use rollers, balls, spirals, domes 3. Go from linear to rotary motion, use centrifugal forces
#25: Self-service	1. Make an object serve itself by performing auxiliary helpful functions 2. Use waste resources, energy, or substances
#35: Parameter changes	1. Change an object's physical state (e.g., to a gas, liquid, or solid) 2. Change the concentration or consistency 3. Change the degree of flexibility 4. Change the temperature
#38: Strong oxidants	1. Replace common air with oxygen-enriched air 2. Replace enriched air with pure oxygen 3. Expose air or oxygen to ionizing radiation 4. Use ionized oxygen. 5. Replace ozonized (or ionized) oxygen with ozone.
#18: Mechanical vibration	1. Cause an object to oscillate or vibrate 2. Increase its frequency (even up to the ultrasonic) 3. Use an object's resonant frequency 4. Use piezoelectric vibrators instead of mechanical ones 5. Use combined ultrasonic and electromagnetic field oscillations
#16: Partial or excessive actions	If 100% of an object is hard to achieve using a given solution method, then by using "slightly less" or "slightly more" of the same method, the problem may be considerably easier to solve
#15: Dynamics	1. Allow (or design) the characteristics of an object, external environment, or process to change to be optimal or to find an optimal operating condition 2. Divide an object into parts capable of movement relative to each other 3. If an object (or process) is rigid or inflexible, make it movable or adaptive
#1: Segmentation	1. Divide an object into independent parts 2. Make an object easy to disassemble 3. Increase the degree of fragmentation or segmentation
#28: Mechanics substitution	1. Replace a mechanical means with a sensory (optical, acoustic, taste or smell) means 2. Use electric, magnetic, and electromagnetic fields to interact with the object 3. Change from static to movable fields, from unstructured fields to those having structure 4. Use fields in conjunction with field-activated (e.g., ferromagnetic) particles
#27: Cheap short-living objects	Replace an inexpensive object with a multiple of inexpensive objects, comprising certain qualities (such as service life).
#24: Intermediary	1. Use an intermediary carrier article or intermediary process 2. Merge one object temporarily with another (which can be easily removed)
#2: Taking out	Separate an interfering part or property from an object, or single out the only necessary part (or property) of an object
#40: Composite materials	Change from uniform to composite (multiple) materials

nature of the inventive principles, not all of the principles were explored for possible solutions to the specific WFH contradictions.

4.3.4.1   Solutions to Employer Perception Issue

#23: Feedback—Recognitions: More frequent and significant recognitions could be introduced to employees who suffer from high stress attributed to WFH practices which cause them to be more engaged with work than usual. For instance, while it is acknowledged that the stress level of employees in the banking industry can be very high [45], a study of commercial banks in Kenya proved that job promotion and employee recognition positively and significantly influence employee performance [46]. According to Merino and Privado [47], it is important to develop employee recognition policies in organizations for the impact it has, not only on well-being, but also on the positive psychological functioning of the workers. Thus, recognitions could be one way of reaffirming employees who practice WFH that their efforts are no less valuable than they were from the original work setting.

#25: Self-service—Quality circles: A quality circle includes a team of employees of the same work area who meet regularly to brainstorm ways of resolving issues and improving production in their organization. The use of quality circles has been found effective in enhancing employee job satisfaction and the work–life balance in various stressful working environments [48–49]. Quality circles are beneficial in addressing diversity and operational challenges of an organization, and can considerably contribute to enhanced intra-organizational group cohesiveness [50]. The use of quality circles may be advantageous in alleviating the stress of WFH employees who are frequently engaged with work as it motivates them to participate and resolve issues together, hence creating a sense of "not being alone" in the problem-solving work.

4.3.4.2   Solutions to Challenging Work Schedule Arrangements

#35: Parameter changes—Simplify complex processes: Simplifying complex processes in work often reduces the uncertainty of predictions and improves the efficiency of the work entailed [51–52]. With the amount of time saved from the effectiveness of work, WFH employees would not have to sacrifice their personal and family time.

4.3.4.3   Solutions to Improper Work Environment

#15: Dynamics—Job rotation: If the working environment at home is unsuitable to WFH employees due to household distractions, perhaps a job rotation might help the employee refocus on the tasks required to achieve the goals. For instance, perhaps an employer could rotate the employee from research- and knowledge-oriented jobs which require a lot of concentration to light multitasking jobs which require less concentration. Such a suggestion may help in alleviating the stress from unavoidable distractions at home while allowing the employee to achieve necessary work targets. Job rotations can appease employees and employers via improved skills and knowledge, job satisfaction, job commitment, and identification of strengths for optimized organizational performance [53–55].

Optimal job rotation schedules also improve productivity, ergonomic risk goals, and workforce flexibility [56]. Perhaps if an employee were to be allowed to rotate his/her job from 100% to 50% WFH practice wherever possible, the work–life balance would improve as well. In regard to using specific equipment, job rotation could allow the appointment of different employees to enter the company premises in turns to use the equipment. Such an arrangement spreads out the technical accountability of employees who could then also take turns to practice WFH without worrying about the technical limitations of accessing equipment.

#1: Segmentation—Break into small teams: Hoegl [57] suggested that teamwork improves with smaller teams, and that teams of three work better than teams of six, which, in turn, work better than teams of nine. Regrouping employees into small teams compared to managing them in a large group based on function may be able to alleviate the stress of communication and volatility of expectations. If managed well, the work targets distributed across a number of small teams may also be less overwhelming than targets from a large group of employees. This arrangement makes work efficient, especially for WFH employees who may be distracted by household responsibilities.

4.3.4.4 Solutions to Lack of Face-to-Face Communication

#40: Composite materials—Use systems thinking in all forms of communications: Systems thinking is a way of making sense of the complexity of the world by looking at it in terms of wholes and relationships rather than by splitting it down into its parts [58]. If the lack of face-to-face communications can lead to misunderstandings or miscommunications, perhaps it is important for WFH employees to adopt a more efficient approach such as systems thinking for communication in virtual meetings. For instance, researchers of health systems found that process mapping (a systems thinking approach) was able to improve communication between stakeholders, strategic data-driven decision-making, and ancillary and supportive health system services [59].

4.4 CONCLUSION

This study aimed to propose ergonomic recommendations to potentially overcome WFH challenges using TRIZ. Using several tools of TRIZ such as the technical contradiction, 39 system parameters, and contradiction matrix, a few inventive principles were extracted as solution models to the contradictions. The selected inventive principles and ergonomic recommendations included #23 Feedback (Recognitions), #25 Self-service (Quality circles), #35 Parameter changes (Simplify complex processes), #15 Dynamics (Job rotation), #1 Segmentation (Break into small teams), and #40 Composite materials (Use systems thinking in all forms of communications). The use of any of these solution models or recommendations can potentially resolve WFH issues that included challenges related to employer perception, work schedule arrangements, improper work environment, and lack of face-to-face communication.

In future, researchers could possibly look into methods of testing out these ergonomic recommendations in the form of formal HRM guidelines or policies for feasibility validation, and perhaps for selecting the most effective recommendation. Researchers could also consider trying out other TRIZ tools to approach WFH challenges, such as function modeling, cause-and-effect chain analysis, substance-field modeling, or physical contradiction. Quantitative and qualitative studies could also be carried out among employers and employees with regard to WFH practices. Some of the possible types of quantitative and qualitative research methods could include surveys, interviews, or focus groups.

The ergonomic recommendations proposed in this study can help organizations and workers maintain the positive aspects of WFH practices while reaping the benefits of work–life balance. By using TRIZ to address issues of WFH practices and work–life balance among employees and employers, organizations may in a sense be able to "secure more than enough for Peter and Paul to be happy about".

ACKNOWLEDGMENTS

The researchers gratefully thank the faculties, schools, departments, and universities for their support in allowing this research to be carried out. The researchers also thank the Technology Transfer Office of Multimedia University for granting the public disclosure approval for this paper. Special thanks go to Chiew Fen Ng for her constructive criticism of the manuscript.

REFERENCES

1. Crosbie T, Moore J. Work–Life Balance and Working from Home. *Social Policy and Society.* 2004;3(3):223–233. doi: 10.1017/S1474746404001733.
2. Lund S, Madgavkar A, Manyika J, et al. *The Future of Work After COVID-19 2021* [26 August 2022]. Available from: www.mckinsey.com/featured-insights/future-of-work/the-future-of-work-after-covid-19
3. Heidt L, Gauger F, Pfnür A. Work from Home Success: Agile Work Characteristics and the Mediating Effect of Supportive HRM. *Review of Managerial Science.* 2022. doi: 10.1007/s11846-022-00545-5.
4. Kawaguchi D, Motegi H. Who Can Work from Home? The Roles of Job Tasks and HRM Practices. *Journal of the Japanese and International Economies.* 2021;62:101162. doi: 10.1016/j.jjie.2021.101162.
5. Peters P, Poutsma E, Van der Heijden BIJM, et al. Enjoying New Ways to Work: An HRM-Process Approach to Study Flow. *Human Resource Management* 2014;53(2):271–290. doi: 10.1002/hrm.21588.
6. OSHA. *Act 514: Occupational Safety and Health Act 1994* [26 August 2022]. Available from: www.dosh.gov.my/index.php/legislation/acts-legislation/23-02-occupational-safety-and-health-act-1994-act-514/file.
7. Chen W-S. A TRIZ Approach to Human Resource Management. *International Journal of Systematic Innovation.* 2015;3(3):13–25.
8. Hsu H-T, Tsai B-S, Chen K-T. *A TRIZ Approach to Business Management Formulation—A Case of HRMS Industry.* International MultiConference of Engineers and Computer Scientists, Hong Kong; 13–15 March 2013.
9. Altshuller G. *The Innovation Algorithm: TRIZ, Systematic Innovation and Technical Creativity.* Worcester, MA: Technical Innovation Center; 1999. 1st ed.
10. Yeoh TS. *TRIZ: Systematic Innovation in Business and Management.* Selangor, Malaysia: Firstfruits Publishing; 2014. 1st ed.
11. Dempsey PG. Effectiveness of Ergonomics Interventions to Prevent Musculoskeletal Disorders: Beware of What You Ask. *International Journal of Industrial Ergonomics.* 2007;37(2):169–173. doi: 10.1016/j.ergon.2006.10.009.

12. Bick A, Blandin A, Mertens K. *Work from Home After the COVID-19 Outbreak*. Dallas, TX: Federal Reserve Bank of Dallas; 2020. 1st ed.

13. Carlson DS, Grzywacz JG, Zivnuska S. Is Work—Family Balance More Than Conflict and Enrichment? *Human Relations*. 2009;62(10):1459–1486. doi: 10.1177/00187267 09336500.

14. Halpern DF. Psychology at the Intersection of Work and Family: Recommendations for Employers, Working Families, and Policymakers. *American Psychologist*. 2005;60(5):397–409. doi: 10.1037/0003-066x.60.5.397.

15. Gorgenyi-Hegyes E, Nathan RJ, Fekete-Farkas M. Workplace Health Promotion, Employee Wellbeing and Loyalty During Covid-19 Pandemic—Large Scale Empirical Evidence from Hungary. *Economies*. 2021;9(2):55. PubMed PMID. doi:10.3390/ economies9020055.

16. Apollo Technical. *Surprising Working from Home Productivity Statistics* [26 August 2022]. Available from: www.apollotechnical.com/working-from-home-productivity-statistics/

17. Adisa TA, Antonacopoulou E, Beauregard TA, et al. Exploring the Impact of COVID-19 on Employees' Boundary Management and Work–Life Balance. *British Journal of Management*. 2022. doi: 10.1111/1467-8551.12643.

18. Dockery AM, Bawa S. Is Working from Home Good Work or Bad Work? Evidence from Australian Employees. *Australian Journal of Labour Economics*. 2014;17(2):163–190.

19. Bernama. *Ugly Side Unfolding About Working from Home 2021* [26 August 2022]. Available from: www.nst.com.my/news/nation/2021/06/698384/ugly-side-unfolding-about-working-home.

20. Olson-Buchanan JB, Boswell WR. Blurring Boundaries: Correlates of Integration and Segmentation Between Work and Nonwork. *Journal of Vocational Behavior*. 2006;68(3):432–445. doi: 10.1016/j.jvb.2005.10.006.

21. Waizenegger L, McKenna B, Cai W, et al. An Affordance Perspective of Team Collaboration and Enforced Working from Home During COVID-19. *European Journal of Information Systems*. 2020;29(4):429–442. doi: 10.1080/ 0960085X.2020.1800417.

22. Pluut H, Wonders J. Not Able to Lead a Healthy Life When You Need It the Most: Dual Role of Lifestyle Behaviors in the Association of Blurred Work-Life Boundaries with Well-Being. *Frontiers in Psychology*. 2020;11 [Original Research]. doi: 10.3389/ fpsyg.2020.607294. English.

23. Maurer R. *Remote Employees Are Working Longer Than Before 2020* [26 August 2022]. Available from: www.shrm.org/hr-today/news/hr-news/pages/remote-employees-are-working-longer-than-before.aspx.

24. Bergefurt L, Appel-Meulenbroek R, Maris C, et al. The Influence of Distractions of the Home-Work Environment on Mental Health During the COVID-19 Pandemic. *Ergonomics*. 2022:1–18. doi: 10.1080/00140139.2022.2053590.

25. Galanti T, Guidetti G, Mazzei E, et al. Work from Home During the COVID-19 Outbreak: The Impact on Employees' Remote Work Productivity, Engagement, and Stress. *Journal of Occupational and Environmental Medicine*. 2021;63(7).

26. Leroy S, Schmidt AM, Madjar N. Working from Home During COVID-19: A Study of the Interruption Landscape. *Journal of Applied Psychology*. 2021;106(10):1448–1465. doi: 10.1037/apl0000972.

27. Mustajab D, Bauw A, Rasyid A, et al. Working from Home Phenomenon as an Effort to Prevent COVID-19 Attacks and Its Impacts on Work Productivity. *TIJAB (The International Journal of Applied Business)*. 2020;4(1):13–21. doi: 10.20473/tijab. V4.I1.2020.13-21.

28. Suarlan S. Teleworking for Indonesian Civil Servants: Problems and Actors. *BISNIS & BIROKRASI: Jurnal Ilmu Administrasi dan Organisasi*. 2017;24(2):100–109.

29. Miller C, Rodeghero P, Storey MA, et al., editors. "How Was Your Weekend?" Software Development Teams Working from Home During COVID-19. *2021 IEEE/ACM 43rd International Conference on Software Engineering (ICSE).* 2021:624–636.
30. Teodorovicz T, Sadun R, Kun AL, et al. How Does Working from Home During COVID-19 Affect What Managers Do? Evidence from Time-Use Studies. *Human–Computer Interaction.* 2021:1–26. doi: 10.1080/07370024.2021.1987908.
31. Kramer A, Kramer KZ. The Potential Impact of the Covid-19 Pandemic on Occupational Status, Work from Home, and Occupational Mobility. *Journal of Vocational Behavior.* 2020;119:103442. doi: 10.1016/j.jvb.2020.103442. PubMed PMID: 32390661; PubMed Central PMCID: PMCPMC7205621. English.
32. Hofstede G. Dimensionalizing Cultures: The Hofstede Model in Context. *Online Readings in Psychology and Culture.* 2011;2(1):8.
33. Himawan KK, Helmi J, Fanggidae JP. The Sociocultural Barriers of Work-from-Home Arrangement Due to COVID-19 Pandemic in Asia: Implications and Future Implementation. *Knowledge and Process Management.* 2022;29(2):185–193. doi: 10.1002/kpm.1708.
34. Xiao Y, Becerik-Gerber B, Lucas G, et al. Impacts of Working from Home During COVID-19 Pandemic on Physical and Mental Well-Being of Office Workstation Users. *Journal of Occupational and Environmental Medicine.* 2021;63(3).
35. Lebow JL. The Challenges of COVID-19 for Divorcing and Post-Divorce Families. *Family Process.* 2020;59(3):967–973. doi: 10.1111/famp.12574. PubMed PMID: 32594521; PubMed Central PMCID: PMCPMC7361269. English.
36. de Klerk JJ, Joubert M, Mosca HF. Is Working from Home the New Workplace Panacea? Lessons from the COVID-19 Pandemic for the Future World of Work. *SA Journal of Industrial Psychology.* 2021;47. doi: 10.4102/sajip.v47i0.1883.
37. Ekpanyaskul C, Padungtod C. Occupational Health Problems and Lifestyle Changes Among Novice Working-from-Home Workers Amid the COVID-19 Pandemic. *Safety and Health at Work.* 2021;12(3):384–389. doi: 10.1016/j.shaw.2021.01.010.
38. Guler MA, Guler K, Guneser Gulec M, et al. Working from Home During a Pandemic: Investigation of the Impact of COVID-19 on Employee Health and Productivity. *Journal of Occupational and Environmental Medicine.* 2021;63(9):731–741.
39. Yeow J, Ng P, Lim W. Workplace Ergonomics Problems and Solutions: Working from Home [version 1; peer review: 1 approved with reservations]. *F1000Research.* 2021;10(1025). doi: 10.12688/f1000research.73069.1.
40. Li X-N, Rong B-G, Kraslawski A. Synthesis of Reactor/Separator Networks by the Conflict-Based Analysis Approach. In: Grievink J, van Schijndel J, editors. *Computer Aided Chemical Engineering.* Amsterdam: Elsevier; 2002. Vol. 10. pp. 241–246.
41. Savransky SD. *Engineering of Creativity: Introduction to TRIZ Methodology of Inventive Problem Solving.* Boca Raton, FL: CRC Press; 2000. 1st ed.
42. Yeoh TS, Yeoh TJ, Song CL. *TRIZ: Systematic Innovation in Manufacturing.* Selangor, Malaysia: Firstfruits Publishing; 2015. 1st ed.
43. Childs PRN. 3—Ideation. In: Childs PRN, editor. *Mechanical Design Engineering Handbook.* Oxford: Butterworth-Heinemann; 2019. 2nd ed. pp. 75–144.
44. Moehrle MG. What Is TRIZ? From Conceptual Basics to a Framework for Research. *Creativity and Innovation Management.* 2005;14(1):3–13. doi: 10.1111/j.1476-8691.2005.00320.x.
45. Michailidis M, Georgiou Y. Employee Occupational Stress in Banking. *Work.* 2005;24:123–137.
46. Kathina C, Bula H. Effects of Recognition and Job Promotion on Employee Performance of Commercial Banks in Kenya. *European Journal of Business and Management.* 2021;13(8):47–53.

47. Merino MD, Privado J. Does Employee Recognition Affect Positive Psychological Functioning and Well-Being? *The Spanish Journal of Psychology*. 2015;18:E64. doi: 10.1017/sjp.2015.67.
48. Hosseinabadi R, Karampourian A, Beiranvand S, et al. The Effect of Quality Circles on Job Satisfaction and Quality of Work-Life of Staff in Emergency Medical Services. *International Emergency Nursing*. 2013 Oct;21(4):264–270. doi: 10.1016/j.ienj.2012.10.002. PubMed PMID: 23266112. English.
49. Samarajeewa CT, Rajaratnam D, Disaratna PAPVDS, et al. Quality Circles: An Approach to Determine the Job Satisfaction of Construction Employees. *International Journal of Construction Education and Research*. 2021:1–16. doi: 10.1080/15578771.2021.1950243.
50. Tamunomiebi MD, Okpara EN. Quality Circles and Intra-Organisational Group Cohesiveness. *Kuwait Chapter of the Arabian Journal of Business and Management Review*. 2020;9(1):40–45.
51. Peters DPC, Yao J, Huenneke LF, et al. A Framework and Methods for Simplifying Complex Landscapes to Reduce Uncertainty in Predictions. In: Wu J, Jones KB, Li H, et al., editors. *Scaling and Uncertainty Analysis in Ecology*. Dordrecht: Springer Netherlands; 2006. pp. 131–146.
52. Worren N. *Organization Design: Simplifying Complex Systems*. London: Routledge; 2018. 2nd ed.
53. Casad S. Implications of Job Rotation Literature for Performance Improvement Practitioners. *Performance Improvement Quarterly*. 2012;25(2):27–41. doi: 10.1002/piq.21118.
54. Alfuqaha OA, Al-Hairy SS, Al-Hemsi HA, et al. Job Rotation Approach in Nursing Profession. *Scandinavian Journal of Caring Sciences*. 2021;35(2):659–667. doi: 10.1111/scs.12947.
55. Ho W-H, Chang CS, Shih Y-L, et al. Effects of Job Rotation and Role Stress Among Nurses on Job Satisfaction and Organizational Commitment. *BMC Health Services Research*. 2009;9(1):8. doi: 10.1186/1472-6963-9-8.
56. Mossa G, Boenzi F, Digiesi S, et al. Productivity and Ergonomic Risk in Human Based Production Systems: A Job-Rotation Scheduling Model. *International Journal of Production Economics*. 2016;171:471–477. doi: 10.1016/j.ijpe.2015.06.017.
57. Hoegl M. Smaller Teams–Better Teamwork: How to Keep Project Teams Small. *Business Horizons*. 2005;48(3):209–214. doi: 10.1016/j.bushor.2004.10.013.
58. Ramage M, Shipp K. *Systems Thinkers*. London: Springer; 2009. 1st ed.
59. Durski KN, Naidoo D, Singaravelu S, et al. Systems Thinking for Health Emergencies: Use of Process Mapping During Outbreak Response. *BMJ Global Health*. 2020;5(10):e003901. doi: 10.1136/bmjgh-2020-003901.

5 Impact of Strategic Orientation and Supply Chain Integration on Firm's Innovation Performance
A Mediation Analysis

*Sonia Umair, Umair Waqas, Ibrahim Rashid
Al Shamsi, Hyder Kamran, and Beata Mrugalska*

CONTENTS

DOI: 10.1201/9781003383444-5

5.1 INTRODUCTION

In the evolving environmental conditions of the business, firms are steadily endeavoring for new business opportunities to survive and maintain market share. For satisfactory innovation performance, firms need to focus on the integration of knowledge mechanisms with their supply chain members (Kumar, Jabarzadeh, Jeihouni, & Garza-Reyes, 2020). For this reason, they consistently have to focus on the outside of the firm boundaries to increase the inflow of knowledge (Soto-Acosta, Popa, & Martinez-Conesa, 2018). The idiosyncrasy of the firm's operational activities in the supply chain has recently elevated the research interest of scholars. Also, nowadays because of economic activities, the ability of the firms to be innovative is crucial for survival (Ardito, Messeni Petruzzelli, Dezi, & Castellano, 2018). Notably, persistent coordination and knowledge integration with their customers and suppliers can significantly highlight the firm's innovation strategies (Hegazy & Ghorab, 2014). The evidence from the study of Kumar et al. (2020), integration with other firms, particularly in the supply chain, through exploitative and explorative forms, might give a different innovative outcome. It was observed that an exploitative integration that usually happens with customers and suppliers leads to improved products, but an explorative integration that happens with universities and research centers mostly results in new products (Faems, Van Looy, & Debackere, 2005) and the integration with these members needs high consideration. Therefore, there is a need to study innovation performance effects from the point of view of the supply chain integration (exploitative and explorative).

According to Ardito et al. (2018) and Noble, Sinha, and Kumar (2002), the philosophies of firms to be strategically orientated help them to achieve innovation performance. Most of the literature focuses on the effects of the strategic orientations individually (Mahmoud & Yusif, 2012; Noble et al., 2002; Rauch, Wiklund, Lumpkin, & Frese, 2002). It has been observed that single orientation may be insufficient for innovation performance (Kumar et al., 2020; Wales, Beliaeva, Shirokova, Stettler, & Gupta, 2020). The use of a fusion of strategic orientation might enable firms to perform better (Day & Lichtenstein, 2007; Deutscher, Zapkau, Schwens, Baum, & Kabst, 2016; Grawe, Chen, & Daugherty, 2009). Furthermore, it is highlighted that the number of overarching strategic orientations exist (Wales et al., 2020) and some of these orientations, that have an important role in innovation, are included in market orientation (accentuate market trend and facilitate customers with innovative values) (Pehrsson, 2014), entrepreneurial orientation (an entrepreneurial emphasis with a focus on developing and working on new product development and services) (Wales et al., 2020), or learning orientation (acquiring and converting knowledge to the firm about market trends, customer needs, and competitive advantage) (Mahmoud & Yusif, 2012).

According to Wales et al. (2020), there is a lack of research where the direct effect of strategic orientations on supply chain integration (mainly exploitative and explorative) was investigated. However, the effect of learning orientation on customer and supplier integration was studied previously (Kumar et al., 2020). Also there is a lack of studies that can explore the collective influence of strategic orientation on innovation performance (Wales et al., 2020).

At the core, the main objective of this research was to understand that how do firms achieve innovation performance by being strategically oriented. For this purpose, this research decomposed three essential strategic orientations and tried to understand how these three orientations, including market, entrepreneurial, and learning orientations separately and complimentary, entail ideas for the firm's innovation performance. Previously, the focus was to consider these strategic orientation aspects individually on product innovation and innovation performance of the firms (Kollmann & Stöckmann, 2012; Kumar et al., 2020; Xian, Sambasivan, & Abdullah, 2018).

The present study also contributed to the literature in several other ways. Initially, this work validated the combined effect of strategic orientation on supply chain integration. Then, this study extended the supply chain integration into exploitative and explorative integrations for further assessment. The work was done extensively on other integrations, including customer, supplier, and other stakeholders (Ellinger, Chen, Tian, & Armstrong, 2015; Munir, Sadiq Jajja, Chatha, & Farooq, 2020; Xian et al., 2018), but not mainly on exploitative and explorative integration. An additional contribution was to understand the mediating effect of supply chain integration with its new dimension (exploitative and explorative) on the individual and joint effect of strategic orientation and innovation performance.

The rest of the chapter proceeds as follows. First, the researchers developed a conceptual framework and research hypotheses. Then, they described the research methodology. Next, they presented the analysis, empirical findings, and the discussion of the results. Finally, the chapter concludes by explaining limitations and proposals for future research.

5.2 RESEARCH HYPOTHESES

In this section, we build on the theoretical framework, shown in Figure 5.1, to derive the hypotheses (H1–H4).

5.2.1 EFFECT OF STRATEGIC ORIENTATION ON INNOVATION PERFORMANCE

The idea of a firm's strategic orientation has recently gained the attention of researchers from management, operations management, marketing, and other literature. The strategic orientations are capabilities and decision-making tendencies of the firm,

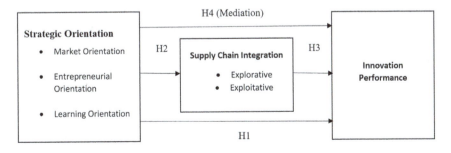

FIGURE 5.1   Research framework.

which lead its activities and provoke behavior, to achieve firm superior performance, develop new products and/services, and determine the ways of strengthening the supply chain (Kumar et al., 2020; Sahi, Gupta, & Cheng, 2020).

Market orientation refers to the firm's activities that generate superior value for the customer. A market-oriented firm looks most efficient and it effectively creates the necessary behavior for the development of superior value for buyers and continuously focuses on the superior performance of the firms (Narver & Slater, 1990). During earlier studies, two main approaches of market orientation were acknowledged (Gupta, Atav, & Dutta, 2017): first is concerned with gathering a wide range of information and dissemination of market information and firm responses (Kohli & Jaworski, 1990) and second is the combination of customer orientation, competitor orientation, and inter-functional coordination (Narver & Slater, 1990). The positive relationship between market orientation and firm performance is also confirmed (Hult & Ketchen, 2001; Mahmoud & Yusif, 2012). A market-oriented firm adjusts its supply chain operations more frequently with its stakeholders, responds immediately to customer demands, and builds a competitive advantage for innovation performance (Deutscher et al., 2016).

It is evident from the previous research (Deutscher et al., 2016; Lumpkin & Dess, 1996; Miller, 2011; Sahi et al., 2020) that in a rapidly changing environment, entrepreneurial orientation–adopted firms are more innovative, risk-takers, and constantly seek out for new opportunities. These firms' psychology constantly focuses on the new product development and services and they embrace a forward-looking and somewhat risky course of action (Lumpkin & Dess, 1996).

Additionally, entrepreneurial orientated firm's inclination remains toward research and innovative ideas, and then commercialization of these into new products and services. For this, they make a substantial investment in research and design (Deutscher et al., 2016; Lumpkin & Dess, 1996; Sahi et al., 2020; Wales et al., 2020). Therefore, firms may benefit from adopting entrepreneurial orientation and increase their performance.

Learning orientation refers to the firms' values and beliefs that lead to the development of knowledge, insight, and awareness (Deutscher et al., 2016; Ellinger et al., 2015; Huber, 1991). The learning process plays an important role in adopting rapid and complex environmental changes. It is evident from the earlier research that the high-learning orientation firms are more competent and can adapt their operational capabilities to respond to the external environment while developing products and services (Lonial & Carter, 2015; Sinkula, Baker, & Noordewier, 1997). Thus, a learning orientation helps firms to create and maintain a competitive position and to enhance innovation performance (Kumar et al., 2020; Wales et al., 2020). In the present study, three dimensions of learning orientation are examined, such as commitment to learning, shared vision, and open-mindedness. The commitment to learning refers to a firm's decision to develop interactive activities (Sinkula et al., 1997) and its substantial effect on investment in training and development. The shared vision focuses on the creation and implementation of knowledge in the organization (Li & Lin, 2006). The open-mindedness refers to evaluating the operational strategies of the firm for the acceptance of new ideas (Baker & Sinkula, 1999).

Prior studies suggest that different combinations of orientations may enable a firm to sustain competitive advantage rather than a single orientation (Boso, Cadogan, & Story, 2012; Ho, Plewa, & Lu, 2016). These results align with the notion of complementarity, which entails a "beneficial interplay of the elements of a system where the presence of one element increases the value of others" (Ennen & Richter, 2010, p. 207). Instead of isolated resources, a complimentary set of resources can also create additional value (Tanriverdi & Venkatraman, 2005). Similarly, every strategic orientation takes a different mechanism to sustain a firm's sales and growth, while the combination of different orientations can create more value as compared to a single one.

To this end, describing a firm's strategic orientation, scholars extensively used a resource-based view to highlight strategic orientation as a rare, valuable, inimitable, and distinctive firm's resources that may work for superior performance (Barney, 1991; Lonial & Carter, 2015).

The extended version of the resource-based view is the dynamic capabilities, in which scholars conceptualize strategic orientation in terms of firm's "capacity (1) to perceive and develop opportunities and threats, (2) to grab opportunities, and (3) to sustain competitiveness through intensifying, combining, protecting, and, when necessary, reconfiguring the business enterprise's intangible and tangible assets" (Teece, 2007, p. 1319). In this vein, with the resource-based view, present research acknowledges that strategic orientation, such as market, entrepreneurial, and learning, positively contributes to increasing a firm's performance (Deutscher et al., 2016). From this, we hypothesized the following relationship:

H1:   Strategic orientation has a positive and significant impact on innovation performance.

5.2.2   EFFECT OF STRATEGIC ORIENTATION ON SUPPLY CHAIN INTEGRATION

Firms interact with several other firms that embrace their supply chain, and integration with them can improve the firm's performance (Munir et al., 2020). The supply chain integration refers to the strategic orientation of the firm with its supply chain partners, who assist firms to integrate external and internal processes, resulting in sustainability in the market, constant flow of information, and smooth operations of the firm in an efficient and effective manner (Wiengarten, Humphreys, Gimenez, & McIvor, 2016; Zhao, Huo, Sun, & Zhao, 2013). Prior studies mostly suggested two types of supply chain integration, famously, internal and external integration (Braunscheidel & Suresh, 2009; Zainol, Abas, & Ariffin, 2016). Internal integration refers to the notion in which firms arrange their suggestions, process, policies, and practices. External integration refers to the coordination with external supply chain partners to manage their processes, planning, and practices (Munir et al., 2020). In the present study, scholars extended supply chain integration to exploitative and explorative integration.

Strategic orientation firm's superior values are to alleviate market, entrepreneurial, and learning skills for product development, transfer of knowledge, and to incorporate with their stakeholders (Deutscher et al., 2016). These firms follow the trends and respond to customer needs and always try to take advantage (Noble et al., 2002). The study of Sahi et al. (2020) argued that strategic orientations are interrelated strategic integration responses to the supply chain uncertainty, and it is evident from the work of Miles and Arnold (1991) that entrepreneurial orientation needs market orientation to target innovative products and to know about the market scenario. More recently, Kumar et al. (2020) suggested that learning orientation is also an important aspect, because learning orientation opens boundaries for exploited and explorative integration. As per the work done by Faems et al. (2005), exploitative and explorative integrations both are required by the supply chains across the organizational boundaries.

A firm aligns explorative and exploitative integration with its internal and external opportunities in an efficient manner and increases performance (Fernhaber & Patel, 2012) and for this it is highlighted that blending the portfolio of explorative and exploitative integration can get valuable opportunities in existing and emerging markets to increase performance.

An exploitative and explorative integration lens delivers a different perspective from which to consider joint criticality of these three strategic orientations. For a long, it is highlighted that entrepreneurial orientation captures the exploratory orientation stressing the bold innovation (Rauch et al., 2002), while market orientation focuses on the exploitative outcomes derived from the market trends (Kohli & Jaworski, 1990). However, learning orientation states that "organizational clue" mergers entrepreneurial and market orientation and it permits firms to effectively manage the flow of information, overcome their problems, and develop successful innovation (Sinkula et al., 1997). Based on the aforementioned discussion, we hypothesized the following relationship.

H2: Strategic orientation has a positive and significant impact on supply chain integration.

5.2.3 EFFECT OF SUPPLY CHAIN INTEGRATION ON INNOVATION PERFORMANCE

Supply chain integration is an influential factor to achieve innovation performance. The knowledge-based view of the firm emphasizes the process of acquiring and deploying valuable knowledge to create a fundamental source of sustainable competitive advantage for the firm (Nonaka, 1991, 1994). Grant (1996) suggested that organizational structural adaptation and integration are required to exploit valuable knowledge. The extant research indicates that knowledge learning and sharing are key factors in creating and sustaining supply chain–related competitive advantage (Cheng, 2011; Spekman, Spear, & Kamauff, 2002). Wowak, Craighead, Ketchen, and Hult's (2013) analysis further confirmed the relevance and importance of exploiting supply chain-related knowledge to achieve superior performance. The source of this knowledge would be suppliers, stakeholders, employees, and others which can be public institutions, universities, and research centers (Jin, Wang, Chen, & Wang, 2015). As noted earlier, supply chain integration can be categorized into exploited and explorative. From the lens of knowledge-based view, exploitation and exploration involve two diverse knowledge creation processes (Floyd & Lane, 2010; Kristal, Huang, & Roth, 2010). If the integration occurs through exploitation, the firm entails internal and existing knowledge, enhances existing processes and techniques in the supply chain of the firms to improve performance. The exploration integration mainly emphasizes the generation of knowledge through public and private institutions, research centers, and universities. Supply chain exploitation and exploration integration not only update existing processes and knowledge but their joint integration further improves their effects (Kristal et al., 2010). He found that explorative and exploitative supply chains both turn to lead a high level of firm performance. The study conducted by Faems et al. (2005) showed that exploitation and exploration collaboration significantly affect the innovation performance of the firms. However, Sahi et al. (2020) emphasized that in the supply chain, there is a need to consider the balanced view of explorative and exploitative between two, which can have a substantial effect on a firm's performance.

Based on the aforementioned discussion, we hypothesized the following relationship:

H3: Supply chain integration has a positive and significant impact on performance.

5.2.4   MEDIATING EFFECT

Supply chain integration is a procedure of redefining and synchronizing knowledge sharing and resources (Droge, Vickery, & Jacobs, 2018) within the operational units of a firm to improve the engagement of supply chain partners effectively (Munir et al., 2020). In the supply chain integration literature, the importance of explorative and exploitative integration as a precursor of interfirm integration has been highlighted by Faems et al. (2005). They suggested that the supply chain must anticipate any interfirm concatenation. However, empirical research at the interaction of strategic orientation and supply chain integration tends to ignore the role of explorative and exploitative integration to enhance innovation performance (Faems et al., 2005; Kumar et al., 2020). Hence, in the present study, researcher argued that the substantial impact of supply chain integration on strategic orientation and firm performance will be strengthened through explorative and exploitative integration. Combining the positive association between explorative and exploitative integration with our first three hypotheses, we proposed the following hypotheses:

H4:   Supply chain integration mediates the relationship between strategic orientation and innovation performance.

5.3   METHODOLOGY

To test our hypothesis, the present study adopted a quantitative-based approach. Survey data was collected through a structured questionnaire. This study collected data from the manufacturing SMEs in the context of surgical instrument manufacturers, which are around 2300 and almost 90% are based in Sialkot, Pakistan (PITAD 2018). That is the reason, this city is known as the hub of surgical manufacturer exporters. To collect data, the questionnaire was distributed to surgical manufacturing SMEs of Sialkot,

Pakistan, through the pick and drop method, obtained personally from the owners of the SMEs in the industrial areas of Sialkot. The important criterion for choosing the SMEs was the number of employees. The organizations with more than 100 employees were considered to maintain sophisticated supply chain management practices.

Once the questionnaire was finalized with the items related to the constructs of the study, three academicians from the management and particularly specialized in SMEs and ten owners of the SMEs had a look at the items to find the relevancy of measurement items and their operationalization. They also checked the wording of the questionnaire and after the suggestions, changes were introduced accordingly.

For distributing the questionnaires, a list of the industries was obtained from the Chamber of Commerce, and we adopted convenience sampling for the collection of the data. The questionnaire included the cover letter with the introduction of the study and it was requested that the questionnaire should be filled by a single respondent including the manager/owner. To retain the standard of the data, questionnaires filled by non-relevant respondents were removed from further processing. Out of the 350 SMEs contacted for data collection, 319 provided complete information. Table 5.1 shows the demographics of the respondents.

5.3.1 Measurement Scale

In this study, the measurement variable consisted of two parts. The first part was of three main sections, each for measuring a particular construct related to the conceptual framework. All the items were measured on a seven-point Likert scale (Likert, 1932).

TABLE 5.1
Demographics

1	Nature of the industry	High technology	30
		Medium technology	230
		Low technology	59
	Total		319
2	Gender (owner)	Male	319
		Female	00
	Total		319
3	Establishment year	Before 1980	59
		1981–2000	240
		2001 onward	20
	Total		319
4	Business Founder is the Manager	Yes	250
		No	69
	Total		319
5	Business Market	Domestic	89
		Foreign	50
		Both	180
	Total		319

Strategic orientation was measured with the three constructs. Market orientation was adapted from the work of Narver and Slater (1990), entrepreneur learning orientation was measured using the work of Covin and Slevin (1989) with the dimensions of innovativeness, proactiveness, and risk-taking. The learning orientation was measured with the dimensions of commitment of learning, shared vision, and open-mindedness, following Sinkula et al. (1997). The firm's supply chain integration was measured using two dimensions adapted from Jansen, Van den Bosch, and Volberda (2005), Smith, Collins, and Clark (2005), Todorova and Durisin (2007), and Marsh and Stock (2006). The first dimension was exploitation which is based on transmutation and application with each containing four items. The other is an exploration that includes recognition and assimilation. Finally, items from innovation performance were adopted from Prajogo and Ahmed (2006) and Prajogo and Sohal (2006). Respondents will compare their performance with their competitors.

5.4   RESULTS

5.4.1   MEASUREMENT MODEL

Standard measures were used for verifying the acceptability of scales. The assessment of measurement items consists of composite reliability for internal consistency, individual indicator reliability, and average variance extracted (AVE) to evaluate convergent validity (Hair, Sarstedt, Hopkins, & Kuppelwieser, 2014). First, the model's internal consistency was assessed. Generally, Cronbach's Alpha is used to estimate reliability based on intercorrelations of the indicators, for which the value should be greater than 0.7 (Hair et al., 2014). In PLS-SEM, internal consistency is also assessed with composite reliability. Composite reliability for all the reflective constructs should be greater than 0.7 (Hair et al., 2014). As shown in Table 5.2, the composite reliability of all the constructs was indeed above 0.7, which shows that the constructs had high internal consistency.

Discriminant validity is used to determine that one construct is empirically distinct from another (Hair, Black, Babin, & Anderson, 2010). For this purpose, the Fornell and Larcker criterion was first used (Fornell & Larcker, 1981). Table 5.3 indicates that the constructs exhibited discriminant validity as the square root of the AVEs were larger than their correlations.

TABLE 5.2
Construct Reliability and Validity

Variables	Alpha	CR	(AVE)
Entrepreneurial orientation	0.911	0.928	0.617
Exploitative	0.744	0.854	0.661
Explorative	0.678	0.861	0.756
IP_	0.720	0.826	0.544
Learning orientation	0.777	0.850	0.540
Market orientation	0.845	0.883	0.522

For further confirmation of discriminant validity, the Heterotrait-Monotrait ratio (HTMT) was also used, where it should be less than 0.90 (Henseler, Hubona, & Ray, 2016). All HTMT values showed sufficient discriminate validity. These results validated that all the items were empirically distinct from each other.

5.4.2 STRUCTURAL MODEL

Bootstrapping technique was used to examine the significance of the hypothesis with a sample size of 319 at a 95% confidence level. Figure 5.2 shows the results of structural model.

TABLE 5.3
Discriminant Validity

Variables	EO	Expl	Expt	IP	LO	MO
Entrepreneurial orientation	0.785					
Exploitative	0.586	0.813				
Explorative	0.254	0.155	0.87			
IP	0.604	0.555	0.318	0.738		
Learning orientation	0.536	0.498	0.302	0.527	0.735	
Market orientation	0.753	0.57	0.279	0.604	0.527	0.723

Abbreviations: EO: entrepreneurial orientation; Expt: exploitative; Expl: explorative; IP: innovative performance, LO: learning orientation; MO: market orientation.

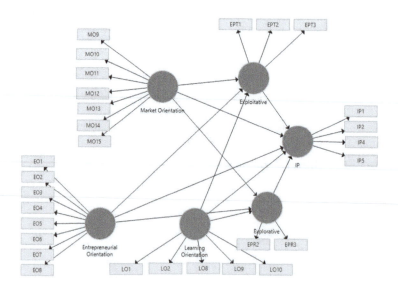

FIGURE 5.2 Structural model results.

TABLE 5.4
Path Coefficients

Hypothesis	Original Sample (O)	Sample Mean (M)	Standard Deviation (STDEV)	T Statistics (|O/STDEV|)	P-Values
Entrepreneurial Orientation -> Exploitative	0.292	0.296	0.079	3.679	0
Entrepreneurial Orientation -> Explorative	0.035	0.034	0.088	0.396	**0.692**
Entrepreneurial Orientation -> IP_	0.198	0.199	0.071	2.805	0.005
Exploitative -> IP_	0.219	0.213	0.059	3.688	0
Explorative -> IP_	0.126	0.128	0.045	2.793	0.005
Learning Orientation -> Exploitative	0.217	0.214	0.063	3.44	0.001
Learning Orientation -> Explorative	0.208	0.214	0.069	3	0.003
Learning Orientation -> IP_	0.163	0.163	0.063	2.606	0.009
Market Orientation -> Exploitative	0.236	0.237	0.084	2.812	0.005
Market Orientation -> Explorative	0.143	0.143	0.081	1.757	**0.079**
Market Orientation -> IP_	0.209	0.213	0.074	2.814	0.005

Note: Significance < 0.05; significance < 0.1.

The results of data analysis show that 9 out of 11 hypotheses are supported as exhibited in Table 5.4.

The impact of three aspects of strategic orientation (entrepreneurial, market, and learning orientation) on the innovation performance of the firm is provided. All three hypotheses are supported and show a positive impact, thus supporting H1/a/b/c. According to the findings, entrepreneurial orientation has positive and significant impact ($\beta = 0.199$, p-value < 0.005), learning orientation ($\beta = 0.163$, p-value < 0.009), and market orientation ($\beta = 0.213$, p-value < 0.005), thus supporting H2/a/b/c/d/e/f.

The results showed supply chain integration (Exploitative and Explorative) to have a positive and significant effect on innovation performance. The results show that exploitative integration ($\beta = 0.213$, p-value < 0.000) and explorative integration ($\beta = 0.128$, p-value < 0.005) have a positive and significant effect on the innovation performance of the firm, thus H3/a/b are supported.

To test the mediation effects, the present study used bootstrapping approach of Preacher and Hayes (2008) with 5000 iterations. The results indicate that supply chain integration (Exploitative) mediates the relation between strategic orientation learning (0.007), entrepreneurial (0.012), and market orientation (0.02) with innovation performance of the firms. Also, the values of the indirect effects show that exploitative integration mediates more effectively as compared to explorative integration. The explorative integration didn't mediate the relationship between strategic orientation learning (0.069), entrepreneurial (0.711), and market orientation (0.159) and innovation performance. The results of indirect effects have been shown in Table 5.5.

The findings of the study also indicate that exploitative integration mediates the relationship between strategic orientation and innovation performance.

TABLE 5.5

Specific Indirect Effect/Mediator

Hypothesis	Original Sample (O)	Sample Mean (M)	Standard Deviation (STDEV)	T Statistics (|O/STDEV|)	P-Values
Learning Orientation -> Exploitative -> IP_	0.047	0.045	0.018	2.712	0.007
Entrepreneurial Orientation -> Exploitative -> IP_	0.064	0.063	0.026	2.499	0.012
Market Orientation -> Exploitative -> IP_	0.052	0.051	0.022	2.324	0.020
Entrepreneurial Orientation -> Explorative -> IP_	0.004	0.005	0.012	0.371	**0.711**
Market Orientation -> Explorative -> IP_	0.018	0.018	0.013	1.408	**0.159**
Learning Orientation -> Explorative -> IP_	0.026	0.028	0.014	1.819	**0.069**

5.5   DISCUSSION

This study attempted to find the impact of strategic orientation on innovation performance. Strategic orientation such as learning, entrepreneurial, and market orientation has a positive and significant effect on innovation performance. The findings of
learning orientation support the previous results (Kumar et al., 2020; Wales et al.,
2020), in which learning orientation showed to be closely related to innovation activities. If an organization focuses on learning orientation, it can adapt its operational
capabilities to respond to the demand of the external environment while developing
and improving products and services (Lonial & Carter, 201). Learning from the environment and the opportunities which may exist, learning orientation–focused organizations can respond immediately to the changing environment. The studies of Baker
and Sinkula (2009) and Keith and Stephen (2006) argued that learning orientation
helps firms to maintain competitive advantage and enhance their performance. It
helps to share knowledge across the organization and promote new ideas. As a result,
the combination of external knowledge and internal knowledge leads to creativity
and innovation organization takes initiatives. Learning orientation also affects innovation performance through several other mechanisms that can explain its indirect
effect (Kumar et al., 2020). It is the kind of desire and direction that an organization
may take, and later this can lead to the development of capabilities in knowledge
and learning (Huang & Wang, 2011). On the contrary, innovation performance is an
important organizational performance indicator that is also under the effect of many
factors. In the literature, there are many intermediate variables between learning orientation and innovation performance. Learning orientation and knowledge creation
can occur within the firm and existing assumptions can be challenged and the result
is better performance in innovation (D'Angelo & Presutti, 2018). The entrepreneurial

orientation has a positive impact on innovation performance. Previous studies also support this argument (Kollmann & Stöckmann, 2012; Sahi et al., 2020; Wales et al., 2020). Through entrepreneurial orientation firm's tendency to experiment and innovate, commercialize them into new products and development that increases the performance of the firms. This study proved that firms who are engaged in entrepreneurial orientation can increase their performance. The study of Covin and Miller (2014) and Covin and Slevin (1989) argued that firms with a high level of entrepreneurial orientation pursue bold, more radical innovation, are risk-oriented, and immediately respond to changing environment of the market. In this study, market orientation affects innovation performance significantly. It may become the cost of competing and source of advantage. Present research findings also support the argument of Narver and Slater (1990) that within the firm after controlling important market-level and business-level influence, market orientation and performance are strongly related. These findings are a clear message to the firms, a substantial market orientation must be the foundation for a business's competitive advantage strategy. Meanwhile, the overall, shared variance among learning, entrepreneurial, and market orientation appears to increase the firm innovation performance.

This study extends supply chain integration in two dimensions (exploitative and explorative) and supports with testing the relationship of these with strategic orientation. The effect of strategic orientation on supply chain integration (exploitative) is statistically significant. The present study supports the hypothesis H2/a/b/c. As stated previously, supply chain integration with two dimensions (i.e., internal and external) helps learning-oriented firms to identify and assimilate new knowledge and open their platform for external and internal knowledge (Kumar et al., 2020). If the concept of supply chain integration, which means improvement, refinement, efficiency, selection, and implementation of tacit knowledge from the stakeholders, is taken into account, the present research concludes that exploitative integration is immune in case of having true strategic orientation as the strategic-oriented firm is committed to innovative performance. In addition, learning orientation develops new ideas and knowledge through supply chain exploitative, while market-oriented firms are more determined on supply chain integration exploitative outcomes to extract market trends. The effect of entrepreneurial and market orientation with other dimensions of supply chain integration (explorative) is not statistically proven in this study, which is contrary to other studies (Narver & Slater, 1990; Sahi et al., 2020). However, learning orientation has a significant effect on explorative integration. In the unit of analysis of this study, there is a lack of collaboration with universities and research institutes; this is the major reason for not supporting this relationship. The study of Memon, Qureshi, and Jokhio (2020) highlighted that "the inability to adapt to new technology, a lack of awareness of the benefits of effective knowledge management, deficiencies in formal language and employee empowerment are among the main obstacles to knowledge creation and sharing." Our third hypothesis H3/a/b was statistically and significantly supported in this study. The supply chain integration impacts positively and significantly on the innovation performance. The effect of collaboration with stakeholders and research institution on innovation performance was shown in the previous literature (Baum, Calabrese, & Silverman, 2000; Rogers, 2004). Several partners such as customers and suppliers, competitors, universities, and research centers are among those with whom collaboration can improve innovative capabilities of the firm (Faems et al., 2005). The reason can be the access to complementary assets required for commercializing innovative ideas successfully, sharing costs among different parties, and coordinating with the stakeholders that reduce risks for one single firm (Ellegaard, 2008). The collaboration about knowledge sharing is important factor and promote innovation (Scala & Lindsay, 2021). Integrating with stakeholders, particularly in supply chain of firms, assist the inflow of knowledge and can increase the firm innovation performance. In their study, Griffin and Hauser (1996) mentioned that with a high level of supply chain integration with their stakeholders, the firm can easily increase the ideas, feedback, and it will help for new product development and process run smoothly. It will also help to increase the probability of the success of new product and better performance in innovation.

5.6   IMPLICATIONS

5.6.1   THEORETICAL IMPLICATIONS

Supply chain integration is one of the pivotal variables in this study with two new dimensions: exploitative and explorative that were not studied before, and are included in a new conceptual framework. The other important element is three strategic oriented variables (market, entrepreneurial, and learning) lead to conduct the relationship with supply chain integration and the effect on innovation performance. Although some studies were conducted to understand the effect of supply chain–related mechanisms on innovation performance, the direct effect of strategic orientation on innovation performance was yet to be studied before and this is one of the major contributions in the literature. Also, the effect of strategic orientation on innovation performance is studied from a supply chain integration perspective. The effect of strategic orientation with three famous variables, i.e., market, learning, and entrepreneurial on supply chain integration in terms of exploitative and explorative that contribute to the development of supply chain innovation performance literature, is one of the major contributions of the study.

5.6.2   PRACTICAL IMPLICATIONS

This study has some implications for practitioners as well. For instance, our study focuses on deeper insight in terms of knowledge that would be helpful for a firm to perusing strategic orientation.

A strategic orientation directly and indirectly (Exploitative) has been shown to influence positively on innovation performance. Taking initiatives to promote it throughout the firm will have a significant influence to enhance firm performance. Market orientation should be embedded in the firm's activities that develop strong customer focus, integration marketing across the organization enhances the performance of the firm. The commitment learning–oriented firm's focus on the research and development, training and development, and knowledge management within the organization lead to enhance the performance innovation. Also, the firm's culture and policies have things to discuss on an open platform. There should be initiatives to discuss cross-cultural ideas, knowledge sharing, and knowledge management mechanisms across the firm to absorb the ideas and new product development efficiently. Furthermore, entrepreneurial orientation in the firm can be used to initiate innovativeness and risk-taking activities proactively. This reflects the level of firm growth strategies and identifies exploitative market opportunities.

Supply chain integration is another important factor in the present research that influences innovation performance. The firms can take exploitative and explorative forms, and each has different innovative outcomes. Exploitative integration which usually happens with the customers and suppliers leads to improved products and identify untapped products in the market. The firm with strong explorative integration collaborates with the universities and research centers, for innovation and new products. Mostly, this happens in research-oriented and progressive countries.

5.7 LIMITATION AND FUTURE RESEARCH

The present research has certain limitations and will make suggestions for future research. It refers to a survey in only one city of Pakistan. Therefore, future studies could have a bigger sample size with more cities that can be included in the survey. Also, there were no females in this study which can be a possible limitation and should be considered in other cities. A comparison of the developed and developing economies can also provide interesting results for later studies.

DATA AVAILABILITY STATEMENT

The data that support the findings of this book/chapter are available from the corresponding author (Umair Waqas) upon reasonable request.

REFERENCES

Ardito, L., Messeni Petruzzelli, A., Dezi, L., & Castellano, S. (2018). The influence of inbound open innovation on ambidexterity performance: Does it pay to source knowledge from supply chain stakeholders? *Journal of Business Research*, *1*, April. https://doi.org/10.1016/j.jbusres.2018.12.043

Baker, W. E., & Sinkula, J. M. (2009). The complementary effects of market orientation and entrepreneurial orientation on profitability in small businesses. Journal of small business management, 47(4), 443-464.

Baker, W. E., & Sinkula, J. M. (1999). The synergistic effect of market orientation and learning orientation on organizational performance. *Academy of Marketing Science: Journal*, *27*(4), 411–427. https://doi.org/10.1177/0092070399274002

Barney, J. (1991). Firm resources and sustained competitive advantage. *Journal of Management*, *17*(1), 99–120. https://doi.org/10.1177/014920639101700108

Baum, J.A., Calabrese, T. and Silverman, B.S. (2000), "Don't go it alone: alliance network composition and startups' performance in Canadian biotechnology", Strategic Management Journal, Vol. 21 No. 3, pp. 267-294.

Boso, N., Cadogan, J. W., & Story, V. M. (2012). Complementary effect of entrepreneurial and market orientations on export new product success under differing levels of competitive intensity and financial capital. *International Business Review*, *21*(4), 667–681. https://doi.org/10.1016/j.ibusrev.2011.07.009

Braunscheidel, M. J., & Suresh, N. C. (2009). The organizational antecedents of a firm's supply chain agility for risk mitigation and response. *Journal of Operations Management*, *27*(2), 119–140. https://doi.org/10.1016/j.jom.2008.09.006

Cheng, J. H. (2011). Inter-organizational relationships and knowledge sharing in green supply chains-moderating by relational benefits and guanxi. *Transportation Research Part E: Logistics and Transportation Review*, *47*(6), 837–849. https://doi.org/10.1016/j.tre.2010.12.008

Covin, J. G., & Miller, D. (2014). International entrepreneurial orientation: Conceptual considerations, research themes, measurement issues, and future research directions. *Entrepreneurship: Theory and Practice*, *38*(1), 11–44. https://doi.org/10.1111/etap.12027

Covin, J. G., & Slevin, D. P. (1989). Strategic management of small firms in hostile and benign environments. *Strategic Management Journal*, *10*(1), 75–87.

Day, M., & Lichtenstein, S. (2007). Strategic supply management: The relationship between supply management practices, strategic orientation and their impact on organisational performance. *Journal of Purchasing and Supply Management*, *12*(6 spec. iss.), 313–321. https://doi.org/10.1016/j.pursup.2007.01.005

D'Angelo, A. and Presutti, M. (2018), "SMEs international growth: the moderating role of experience on entrepreneurial and learning orientations", InternationalBusiness Review.

Deutscher, F., Zapkau, F. B., Schwens, C., Baum, M., & Kabst, R. (2016). Strategic orientations and performance: A configurational perspective. *Journal of Business Research*, *69*(2), 849–861. https://doi.org/10.1016/j.jbusres.2015.07.005

Droge, C., Vickery, S. K., & Jacobs, M. A. (2018). Does supply chain integration mediate the relationships between product/process strategy and service performance ? An empirical study. *International Journal of Production Economics*, *137*(2), 250–262. https://doi.org/10.1016/j.ijpe.2012.02.005

Ellegaard, C. (2008). Supply risk management in a small company perspective. *Supply Chain Management: An International Journal*, *13*(6), 425–434. https://doi.org/10.1108/13598540810905688

Ellinger, A. E., Chen, H., Tian, Y., & Armstrong, C. (2015). Learning orientation, integration, and supply chain risk management in Chinese manufacturing firms. *International Journal of Logistics Research and Applications*, *18*(6), 476–493. https://doi.org/10.1080/13675567.2015.1005008

Ennen, E., & Richter, A. (2010). The whole is more than the sum of its parts — or is it? A review of the empirical literature on complementarities in organizations. Journal of Management, 36(1), 207–233. https://doi.org/10.1177/0149206309350083.

Faems, D., Van Looy, B., & Debackere, K. (2005). *The role of inter-organizational collaboration within innovation strategies: Towards a portfolio approach.* DTEW Research Report 0354, 1–33. Retrieved from https://lirias.kuleuven.be/bitstream/123456789/118280/1/OR_0354.pdf

Fernhaber, S. A., & Patel, P. C. (2012). How do young firms manage product portfolio complexity? The role of absorptive capacity and ambidexterity. *Strategic Management Journal*, *33*(13), 1516–1539. https://doi.org/10.1002/smj.1994

Floyd, S. W., & Lane, P. J. (2010). Strategizing throughout the organization: Managing role conflict in strategic renewal. *Academy of Management Review*, *25*(1), 154–177. https://doi.org/10.1057/9780230305335

Fornell, C. & Larcker, D. (1981). Evaluating structural equation models with unobservable variables and measurement error. Journal of Marketing Research, 18(1), 39–50.

Griffin, A. and Hauser, J.R. (1996), "Integrating R&D and marketing: a review and analysis of the literature", Journal of Product Innovation Management: An International Publication ofthe Product Development&Management Association, Vol. 13 No. 3, pp. 191-215.

Grant, R. (1996). Toward a knowledge-based theory of the firm. *Strategic Management Journal*, *17*(S2), 109–122. https://doi.org/10.1002/smj.4250171110

Grawe, S. J., Chen, H., & Daugherty, P. J. (2009). The relationship between strategic orientation, service innovation, and performance. *International Journal of Physical Distribution and Logistics Management*, *39*(4), 282–300. https://doi.org/10.1108/09600030910962249

Gupta, V. K., Atav, G., & Dutta, D. K. (2017). Market orientation research: A qualitative synthesis and future research agenda. *Review of Managerial Science*, *13*(4), 649–670. https://doi.org/10.1007/s11846-017-0262-z

Jansen, J. J. P., Van den Bosch, F. A., & Volberda, H. W. (2005). Managing potential and realised absorptive capacity: How do organizational antecedents matter? *Academy of Management Journal*, *48*, 999–1015.

Jin, X., Wang, J., Chen, S. and Wang, T. (2015), "A study of the relationship between the knowledge base and the innovation performance under the organizational slack regulating", Management Decision, Vol. 53 No. 10, pp. 2202-2225.

Hair, J., Black, W., Babin, B., & Anderson, R. (2010). Multivariate data analysis. London: Prentice Hall.

Hair Jr, J. F., Sarstedt, M., Hopkins, L., & Kuppelwieser, V. G. (2014). Partial least squares structural equation modeling (PLS-SEM): An emerging tool in business research. European business review.

Hegazy, F. M., & Ghorab, K. E. (2014). The influence of knowledge management on organizational business processes' and employees' benefits. *International Journal of Business and Social Science*, *5*(1), 148–172. https://doi.org/10.5171/2015.928262

Henseler, J., Hubona, G., & Ray, P. A. (2016). Using PLS path modeling in new technology research: Updated guidelines. Industrial Management and Data Systems, 116(1), 2–20. https://doi.org/10.1108/IMDS-09-2015-0382

Ho, J., Plewa, C., & Lu, V. N. (2016). Examining strategic orientation complementarity using multiple regression analysis and fuzzy set QCA. *Journal of Business Research*, *69*(6), 2199–2205. https://doi.org/10.1016/j.jbusres.2015.12.030

Huang, S.K. and Wang, Y.L. (2011), "Entrepreneurial orientation, learning orientation, and innovation in small and medium enterprises", Procedia Social and Behavioral Sciences, Vol. 24, pp. 563-570

Huber, G. P. (1991). Organizational learning: The contributing processes and the literatures. *Organization Science*, *2*(1), 88–115. https://doi.org/10.1287/orsc.2.1.88

Hult, G. T. M., & Ketchen, D. J. (2001). Does market orientation matter?: A test of the relationship between positional advantage and performance. *Strategic Management Journal*, *22*(9), 899–906. https://doi.org/10.1002/smj.197

Keith, T., & Stephen, A. (2006). The learning organisation: A meta-analysis of themes in literature. The Learning Organization, 13(2), 123–139. https://doi.org/10.1108/0969647 0610645467.

Kohli, A. K., & Jaworski, B. J. (1990). Market orientation: The construct, research propositions, and managerial implications. *Journal of Marketing*, *54*, 1–18, April.

Kollmann, T., & Stöckmann, C. (2012). Filling the entrepreneurial orientation-performance gap: The mediating effects of exploratory and exploitative innovations. *Entrepreneurship: Theory and Practice*, *38*(5), 1001–1026. https://doi.org/10.1111/j.1540-6520.2012.00530.x

Kristal, M. M., Huang, X., & Roth, A. V. (2010). The effect of an ambidextrous supply chain strategy on combinative competitive capabilities and business performance. *Journal of Operations Management*, *28*(5), 415–429. https://doi.org/10.1016/j.jom.2009.12.002

Kumar, V., Jabarzadeh, Y., Jeihouni, P., & Garza-Reyes, J. A. (2020). Learning orientation and innovation performance: The mediating role of operations strategy and supply chain integration. *Supply Chain Management*, *4*, 457–474, December. https://doi.org/10.1108/SCM-05-2019-0209

Li, S., & Lin, B. (2006). Accessing information sharing and information quality in supply chain management. *Decision Support Systems, 42*(3), 1641–1656. https://doi.org/10.1016/j.dss.2006.02.011

Likert, R. (1932). A technique for the measurement of attitudes. *Archives of Psychology, 22,* 55. https://doi.org/2731047

Lonial, S. C., & Carter, R. E. (2015). The impact of organizational orientations on medium and small firm performance: A resource-based perspective. *Journal of Small Business Management, 53*(1), 94–113. https://doi.org/10.1111/jsbm.12054

Lumpkin, G. T., & Dess, G. G. (1996). Clarifying the entrepreneurial orientation construct and linking it to performance. *Academy of Management Journal, 21*(1), 135–172.

Mahmoud, M. A., & Yusif, B. (2012). Market orientation, learning orientation, and the performance of nonprofit organisations (NPOs). *International Journal of Productivity and Performance Management,61*(6),624–652.https://doi.org/10.1108/17410401211249193

Marsh, S. J., & Stock, G. N. (2006). Creating dynamic capabilities: The role of intertemporal integration, knowledge retention and interpretation. *Journal of Product Innovation Management, 23,* 422–436.

Memon, S. B., Qureshi, J. A., & Jokhio, I. A. (2020). The role of organizational culture in knowledge sharing and transfer in Pakistani banks: A qualitative study. *Global Business and Organizational Excellence, 39*(3), 45–54. https://doi.org/10.1002/joe.21997

Miles, M.P., Arnold, D.R., 1991. The relationship between marketing orientation and entrepreneurial orientation. Entrep. Theory Pract. 15 (4), 49–66.

Miller, D. (2011). Miller (1983) revisited: A reflection on EO research and some suggestions for the future. *Entrepreneurship: Theory and Practice, 35*(5), 873–894. https://doi.org/10.1111/j.1540-6520.2011.00457.x

Munir, M., Sadiq Jajja, M. S., Chatha, K. A., & Farooq, S. (2020). Supply chain risk management and operational performance: The enabling role of supply chain integration. *International Journal of Production Economics,* 1–62. https://doi.org/10.1016/j.ijpe.2020.107667

Narver, J. C., & Slater, S. F. (1990). The effect of a market orientation on business profitability. *Journal of Marketing, 4*(54), 20–35. https://doi.org/10.1016/0737-6782(91)90038-z

Noble, C. H., Sinha, R. K., & Kumar, A. (2002). Market orientation and alternative strategic orientations: A longitudinal assessment of performance implications. *Journal of Marketing, 66*(4), 25–39. https://doi.org/10.1509/jmkg.66.4.25.18513

Nonaka, I. (1991) 'The knowledge-company', Harvard Business Review, Vol. 69, No. 6, pp.96–104.

Nonaka, I. (1994) 'A dynamic theory of organizational knowledge creation', Organization Science, Vol. 5, No. 1, pp.14–37

Pehrsson, A. (2014). Firms' customer responsiveness and performance: The moderating roles of dyadic competition and firm's age. *Journal of Business & Industrial Marketing, 29*(1), 34–44. https://doi.org/10.1108/JBIM-01-2011-0004

Prajogo, D. I., & Ahmed, P. K. (2006). Relationships between innovation stimulus, innovation capacity, and innovation performance. R&D Management, 36(5), 499-515.

Prajogo, D. I., & Sohal, A. S. (2006). The integration of TQM and technology/R&D management in determining quality and innovation performance. *Omega,* 34(3), 296–312.

Preacher, K. J., & Hayes, A. F. (2008). Asymptotic and resampling strategies for assessing and comparing indirect effects in multiple mediator models. Behavior research methods, 40(3), 879-891.

Rogers, M. (2004), "Networks, firm size and innovation", SmallBusiness Economics, Vol. 22 No. 2, pp. 141-153

Rauch Wiklund, J., Lumpkin, G. T., & Frese, M, A. (2002). Entrepreneurial orientation and business performance: An assessment of past research and suggestions for the future. *Entrepreneurship Theory and Practice, 33*(3), 761–767. https://doi.org/10.1017/CBO9781107415324.004

Sahi, G. K., Gupta, M. C., & Cheng, T. C. E. (2020). The effects of strategic orientation on operational ambidexterity: A study of Indian SMEs in the industry 4.0 era. *International Journal of Production Economics*, *220*, 107395, August. https://doi.org/10.1016/j. ijpe.2019.05.014

Scala, B., & Lindsay, C. F. (2021). Supply chain resilience during pandemic disruption: Evidence from healthcare. *Supply Chain Management*, *26*(6), 672–688. https://doi. org/10.1108/SCM-09-2020-0434

Sinkula, J. M., Baker, W. E., & Noordewier, T. (1997). A framework for market-based organizational learning: Linking values, knowledge, and behavior. *Journal of the Academy of Marketing Science*, *25*(4), 305–318. https://doi.org/10.1177/0092070397254003

Smith, K.G., Collins, C.J., Clark, K.D., 2005. Existing knowledge, knowledge creation capability, and the rate of new-product introduction in high-technology firms. Acad. Manag. J. 48, 346–357.

Soto-Acosta, P., Popa, S., & Martinez-Conesa, I. (2018). Information technology, knowledge management and environmental dynamism as drivers of innovation ambidexterity: A study in SMEs. *Journal of Knowledge Management*. https://doi.org/10.1108/ JKM-10-2017-0448

Spekman, R., Spear, J., & Kamauff, J. (2002). Supply chain competency: Learning as a key component. *SSRN Electronic Journal*, *7*, 41–55. https://doi.org/10.2139/ssrn.282519

Tanriverdi, H., & Venkatraman, N. (2005). Knowledge relatedness and the performance of multibusiness firms. Strategic Management Journal, 26(2), 97–119. https://doi.org/10. 1002/smj.435.

Teece, D. J. (2007). Explicating dynamic capabilities: The nature and microfoundations of (sustainable) enterprise performance. Strategic Management Journal, 28(13), 1319–1350. https://doi.org/10.1002/smj.640.

Todorova, G., & Durisin, B. (2007). Absorptive capacity: Valuing a reconceptualisation. *Academy of Management Review*, *32*, 774–786.

Wales, W., Beliaeva, T., Shirokova, G., Stettler, T. R., & Gupta, V. K. (2020). Orienting toward sales growth? Decomposing the variance attributed to three fundamental organizational strategic orientations. *Journal of Business Research*, *109*, 498–510. https://doi. org/10.1016/j.jbusres.2018.12.019

Wiengarten, F., Humphreys, P., Gimenez, C., & McIvor, R. (2016). Risk, risk management practices, and the success of supply chain integration. *International Journal of Production Economics*, *171*, 361–370. https://doi.org/10.1016/j.ijpe.2015.03.020

Wowak, K. D., Craighead, C. W., Ketchen, D. J., & Hult, G. T. M. (2013). Supply chain knowledge and performance: A meta-analysis. *Decision Sciences*, *44*(5), 843–875. https://doi. org/10.1111/deci.12039

Xian, K. J., Sambasivan, M., & Abdullah, A. R. (2018). Impact of market orientation, learning orientation, and supply chain integration on product innovation. *International Journal of Integrated Supply Management*, *12*(1–2), 69–89. https://doi.org/10.1504/ IJISM.2018.095681

Zainol, M. A., Abas, Z., & Ariffin, A. S. (2016). Supply chain integration and technological innovation for business performance of aquaculture contract farming in Malaysia: A conceptual overview. *International Journal of Supply Chain Management*, *5*(3), 86–90.

Zhao, L., Huo, B., Sun, L., & Zhao, X. (2013). The impact of supply chain risk on supply chain integration and company performance: A global investigation. *Supply Chain Management: An International Journal*, *18*(2), 115–131. https://doi.org/10.1108/13598541311318773

6 The South African Automotive Industry's Competitiveness and Supply Chain Integration Challenges

John M. Ikome, Opeyeolu Timothy Laseinde, and M.G. Kanakana Katumba

CONTENTS

6.1 INTRODUCTION

The supply chain (SC) is a very vital aspect of every organization, ranging from input supplies, work in progress (WIP) along the various production workstation, to the final product and delivery to the customer. As a result of its importance, any deviation might completely disrupt the entire workflow, leading to customer's dissatisfaction and loss of mark shares. According to an industry report published by TISA and MIDC (2003), success and modern technology mastery (Industry 4.0) is always seen as successful economic symbol within the automotive industry and any disruption or constrain within the supply chain can be very detrimental to the entire organization and the economy of the country. Figure 6.1 shows the linking of inputs that brings the product to a customer from variety of raw material and processes right up to the final

DOI: 10.1201/9781003383444-6

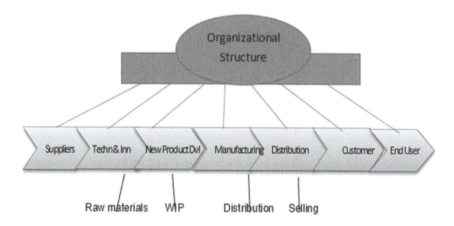

FIGURE 6.1 Organizational structure depicting the linking of inputs to end users from variety of raw material and processes.

Source: Adapted from Lysons, K., & Gillingham, M. (2003). Purchasing and supply chain management. Harlow Prentice Hall.

output. This manufacturing process consists of various number of disintegration and turns to make things a bit complicated.

Hugo, Badenhorst-Weiss, and Van Biljon (2004) define supply chain as the process that encompasses every production effort that is required to produce and deliver a final product from a suppliers' supplier to the customers' customers. Furthermore, Naude and Badenhorst-Weiss (2011a) briefly describe supply chain and how important it is as the competition among companies is narrow but now the era is where the competition is based among different companies supply chains.

A supply chain encompassed a link between two or more parties and the flow of resources including machines, finance, information, and human within the manufacturing process (Naude, and Badenhorst-Weiss, 2011b). In addition, the original companies are no longer manufacturing and assembling product but rather, these products are created as a result of collaboration with many members within the supply chain. In a similar research work, Khayundi (2011) further says streamlined collaboration and effective coordination between suppliers, business partners, and customers can be achieved through efficient supply chain management. This also has become the determinant of company's major business competitive strategy and sustainability model.

6.2   SUPPLY CHAIN MANAGEMENT AND INTEGRATION

The aggressive competition in today's global marketplace, and the introduction of products alongside with shorter life cycles and the increased expectations or constantly changing customer's behavior, has compelled companies to invest in and focus their attention on their supply chains management systems to stay competitive (Wang, Zhang, and Liu, 2021).

According to Caddy and Helou (2007), a company need to differentiate itself from its competitors in the marketplace by enhancing its present in order to gain future profitability through balancing its prices and services levels. This includes delivery of the necessary products at the right time and at the lowest possible cost. The cost rate has driven the creation of cheaper labor and the range of responsibilities has created a co-dependency of supply chain participants.

Co-ordination is crucial because one supply chain member's output is another member's input, so this interdependency commands a fluid, and a seamless flow. A well-integrated supply chain management system can enhance and contribute toward a considerable product quality, shorter lead time, and responsive supply chain, at lower cost, alongside increased levels of customer's satisfaction (Naude and Badenhorst-Weiss, 2011b).

Logistics operation in the automotive supply chain are complex in nature and account for huge expenses and consequently are fragments wherever improvements can be made in order to save and meet customers' requirements (Sturgeon and Van Biesebroeck, 2012). According to Ikome and Kanakana (2018), managing the supply chain today is very critical as it is the central focus for countless businesses. Getting the final product to the customer requires careful planning and considerations of all the aspects of supply chain.

According to Ikome and Kanakana (2018), managing the supply chain today is very critical as it is the central focus for countless businesses. In order to get the final product to the customer, careful planning and considerations of all the aspects of supply chain are required.

Notably, it also includes coordination and collaboration with channel partners, which can be suppliers, intermediaries, third-party service providers, and customers. In essence, supply chain management (SCM) integrates supply and demand management within and across companies (Swink et al., 2011; Council of Supply Chain Management Professional, 2018). Other scholars Bennett and O'Kane (2006), Humphreys et al. (2007), view SCM as a collaborative approach within and across companies that include various role players in the supply chain, which significantly contribute to improved product quality, shorter lead times, and a higher responsiveness of the supply chain, at lower cost and improved customer satisfaction levels.

From these definitions, it is clear that SCM includes all those activities involved in the flow of materials through the supply chain that SCM extends from the ultimate customer back to Mother Earth, and that there is some kind of relationship, collaboration, or cooperation between supply chain parties.

Therefore, SCM in the manufacturing of motorcars can be regarded as the inputs needed at each level or tier of the supply chain, to be transformed into finished automotive components. This Inputs include customers' information, equipment, machinery, raw materials, finances and labor.

The processing and transformation is the management of the modification process that converts inputs (such as raw materials and labor) into outputs (finished goods and services) that are distributed to the customer at the next lower level of the supply chain. This process is continued at all levels of the supply chain until the end product (the motorcar) reaches the final customer. Furthermore, in order for a firm to achieve this end product (the motorcar), the internal and external supply chain (SCM) of the firm cannot be slow down or halt or have a glitch.

6.3 THE AUTOMOTIVE INDUSTRY DEVELOPMENTS

Original equipment manufacturers, automotive component manufacturers, the after-market, and automotive rail are the major parties within the supply chain (SCM) of the automotive industry. The original equipment manufacturer (OEM) is the main player within the supply chain and every other supply chain (SCM) revolved around it. Figure 6.2. depicts this relationship.

According to Naude and Badenhorst-Weiss (2011), due to the constant competitive pressure, industries are faced with constraints and have to look for alternative means to minimize their operational cycle time and cost and improve quality, forecasting techniques, and also persistent emphasis on time base competition, lean manufacturing, and flexibility, particularly the automotive firms which in return have exclusively chosen the designing of vehicles and assembling while on the other end outsourcing the production of automotive components.

A majority of OEMs have invested huge amount of money in order to meet local customers demand by minimizing the production cost of vehicles through the establishment of assembling plants out of their home-based countries (Naude and Badenhorst-Weiss, 2011). As time goes on, the trend of the automotive industry has risen to outsourcing design and manufacturing activities to other suppliers. According to Humphrey and Salerno (2000), first-tier suppliers have taken full responsibilities for manufacturing and supplying systems, design, and modules for the global automotive industry's assemblers. Arnold and Chapman (2004), including Choi and Krause (2006), acknowledge that suppliers are very import to businesses as they fully rely on them for quality enhancement, cost reduction, and the speed of innovation development in order to gain competitive advantage over their competitors.

Without any doubt, this justifies the fact that there is an increase dependency on suppliers of automotive industry due to the increased outsourcing, creating a greater response, responsibility, and trustworthy partnership of suppliers. This therefore

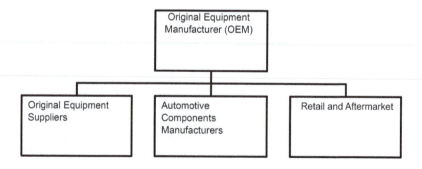

FIGURE 6.2   Industry relationships.

means suppliers have to constantly adopt a continuous improvement strategy in order to be competent and following the trends of frequently demand changes.

According to Arnold and Chapman (2004), the request of Just in Time (JIT) strategy from manufacturing industries has created a change within the business environment, forcing every manufacturing industry and supplier to apply different approach to relationship between firms. Business success depends greatly on the way an organization satisfies its customers' needs or requirements, and with the growing trend on outsourcing, the efficient function of an organization's supply chain (SCM) requires a lot of planning and coordinating activities.

6.4 AN OVERVIEW OF GLOBAL AUTOMOTIVE SUPPLY CHAIN MANAGEMENT

The success of organizations now solely depend on its supply chain management (SCM), and the increase in demand and volatility level requires an organization to respond on time in order to stay within the competitive network, particularly the automotive (Fawcett, Ellram and Ogden, 2007). Ikome and Laseinde (2020) further assented to this statement by saying the supply chain management (SCM) is a very powerful tool that affords a huge number of the opportunity for manufacturing industries to gain competitive advantage.

Supply chain management (SCM) is a system approach that manages an entire flow of customer's requirement, service and materials from input of suppliers, and manufacturing right to the end user (Leenders and Fearon, 2004). A survey done by Gansler, Luby, and Kornberg (2004) noted that supply chain management (SCM) is the management and controlling of all activities ranging from funds, materials, and related information in the logistics process from the acquisition of raw materials to the delivery of finished products to the end user.

SCM today is a fluid process that needs to flow smoothly without any disruption and the growing trend in globalization today, created a great number of advantages and equally risk factors, putting supply chain managers in complex and difficult situations that are difficult to overcome at times (Updike, 2012). Duarte and Machado (2011) equally said supply chain management is very critical to the success of any business or organization nowadays and in addition, it is not merely a challenging factor in elusive goals, and likewise, the manner in which an organization manages the challenges in its supply chain, its goals, improved efficiency and reliability is of great value to its customers.

The automotive industry has experienced great competition in the global market of late due to globalization (Ikome, Laseinde, and Katumba, 2021). Pires and Cardoza (2007) mentioned that it has been facing emerged pressure such as achieving environment-friendly design, response and product delivery time, price reduction, customers service, and quality improvement.

According to Fawcett, Ellram, and Ogden (2007), the ultimate goal of a supply chain management practice is to create an agreement between shareholders and customers, called a value chain delivery system. Teamwork and advanced technology can be used to form efficient and effective goals that create value for the end user or customer.

6.4.1 International Automotive Value Chain

The coordination of all the activities related to the production of goods and service are all related to supply chain management, but slightly differs in relationship to a value chain concept. Morrison, Pietrobelli, and Rabellotti (2008) states that according to Michael Porter, it is the range of production activities that are required from a conceptualized stage, manufacturing, customers supplies and delivery, and then disposal after final use.

This was further acknowledged by Lysons and Gillingham (2003), affirming that value chain is a link to supply chain. Value chain includes all the various processes and activity related to information, supplier, design and manufacturing, marketing and delivering the end product to the customers. These activities can be done in-house or fragmented and divided to firms in a number of locations. Please see value chain in Figure 6.3.

6.4.2 Global Value Chain

A global value chain is a chain of activities which are separated among numerous organizations in various geographical locations in the world. Global value chains covers a full range of interconnected production process and activities that are performed by firms in various geographic locations in order to produce a product or service from input to manufacturing and delivering the final output to the customer UNCTAD, United Nation, (2010).

A research work done by Humphrey and Schmitz (2000) outlined key findings of different investigations to express that access to technologically advanced country markets has facilitated more and more reliance on entering into the global production network of lead firms situated in technologically advanced nations. The analysis of these scholars shows that global value chain is usually a variety of choices for local manufacturing industry and suppliers to gain access to advanced markets and technologies.

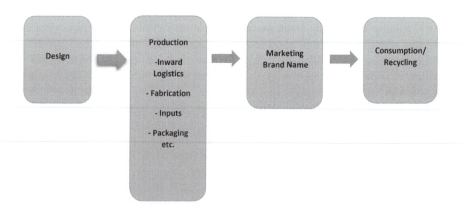

FIGURE 6.3   A value chain.

Source: Adapted from Kaplinsky, Morris, and Readman (2012).

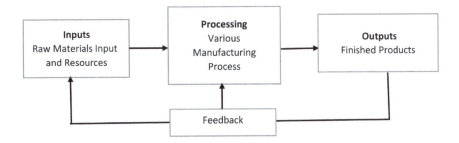

FIGURE 6.4   The supply chain/manufacturing process.

6.4.3   THE SOUTH AFRICAN AUTOMOTIVE SUPPLY CHAIN CASE STUDY

An automotive component manufacturing industry in South Africa was selected for a case study and data collection. Due to confidentiality, the name of the company will not be mentioned.

This company manufactures automotive components that are very critical components to the original equipment manufacturers (OEMs) and as a result any delay within the supply chain can render the entire production process and the end product (outputs) absolute. The primary objective was to study the production process from the initial point of input, transformation, or production process and output, as illustrated in Figure 6.4.

6.5   AUTOMOTIVE SUPPLY CHAIN DISRUPTIONS

According to Ikome, Laseinde and Kanakana (2022), if one can actually measure what he is talking about and express it in numbers, then it means he has clear information about it, and "You cannot manage what you cannot measure". From an engineering point of view, these statements actually demonstrate why measurements are important, yet it is surprising that South African Automotive Industry often overlook this function.

From a manufacturing perspective, the production process includes all activities from raw material inputs to processing, output, and shipment to the customers including feedback, as shown in Figure 6.4. In linking this to the case study automotive industry, it means all the materials that are required to manufacture complete vehicles are classified under input while the processing includes machines and technology. Furthermore, this process has to function as a complete system in order to achieve the objectives and goals of the organization and any deviation to it will be very detrimental to the entire organization.

A research conducted by Ikome, Ayodeji, and Kanakana (2015) reflects that in most cases, senior managements tend to focus more on measurable performance indicators because of the financial implications and reflections. Measuring industries' performance is important, but it is also equally important to measure the implications of disruptions. Unfortunately, many top executives are not comfortable or familiar with disruptions metrics to know how to assess the impact of all these potential disruptions.

The ability of organizations to measure and track the impact of random disruptions, as well as changes in trends over time are important tools to effectively manage and control supply chain disruptions.

It must be emphasized that supply chain disruptions include all potential disruptive factors from receipt of an order to order shipment—this broad process is referred to as a "value chain".

This indicates that when a situation like this occurs, either the productivity output reduces or the production lines come to a complete stand still. Therefore, addressing disruption problems in a manufacturing industry is very important, particularly the South African Automotive industry, where decision-makers and leaders have to manage critical services in the presence of disruptions as manufacturing disturbance is also very detrimental to the competitiveness of an organization.

6.6   RESULTS AND DISCUSSION

According to the results, it shows that both power failure, ports delays, and customers cancellations appear to be significant obstacles to the supply chain challenges in South African automotive industry. Skills development can only be rectified over time, putting it in the hands of manufacturers a little more than labor issues. The results in Table 6.1. and Figure 6.6. reveal that some of the problems facing the South African automobile industry are complicated and difficult to address.

TABLE 6.1

Various Challenges within the South African Automotive Industry

SA Automotive Supply Chain Challenges

Challenges	% of Identified Challenges	% Overwhelming Challenge
Infrastructural challenges		
Port delays	93.1	100
Increased road freight volumes	83.5	98.2
Production/skills challenges		
Labor problems	67.8	77.8
Lack of skills	88.2	69.5
Market/service challenges		
Customers sometimes cancel their orders	99.1	87.1
Difficult to find new markets	58.4	87.2
Improving our service levels	73.8	82.3
Cost challenges		
High operating costs	98.2	83.4
High prices of materials/components	83.8	72.5
High fuel costs affecting operating costs	66.7	59.6
High costs at South African ports	61.2	82.7
Technological challenges		
Power failure	92.1	91.2

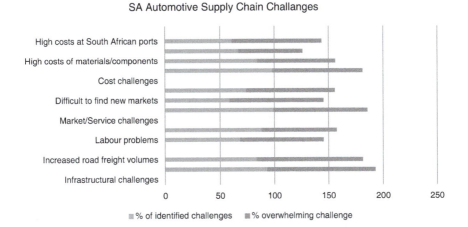

FIGURE 6.5   Level of challenges facing South African automotive.

In terms of technological obstacles, lowering the high costs of replacing outmoded production tools is particularly a significant problem. Infrastructure challenges of high amounts of road freight, including high operational cost resulting from power failure and work in progress (WIP), are also tough to overcome. Because the government (Transnet) is in charge of infrastructure development, manufacturers have little or no control over these issues.

The South African manufacturing industries are dominated by transformation process as a transitional element for change since the inception of democracy in 1994. Numerous studies have been conducted that reveal the lack of competitive models which is the same situation experienced in the South African automotive industry (Council of Supply Chain Management Professional, 2018; Bennett and O'Kane 2006; Humphreys et al., 2007). A survey was conducted by the primary author of this paper in the South Africa automotive components manufacturing industry. The respondent's feedback indicates that automotive industries component manufacturers experience problems related to high fuel expenses, high operating costs, high expenditures incurred at ports, and high material prices. The survey shows that these problems are somewhat difficult, if not impossible, to overcome. This is understandable, as manufacturers are not the major price determinants; the government and the economy's competitive position have an impact on cost dynamics.

Finding new markets was also regarded to a certain degree extremely challenging according to the respondents. The competitive position of the local automotive industry versus foreign car manufacturers, as well as the role of major firms, has influence on these decisions. It is also tough for South African automotive producers due to the impact of rising economies to explore new markets abroad.

As a result, it is recommended that the industry work to gain a competitive advantage in the continent (African markets). Furthermore, the majority of those who agreed that customers occasionally abandon orders said it is a difficult barrier to overcome. This is due to the global chain, as the customers demand changes constantly due to the in-flogs of cheap product from China and other nations. Skills shortages have been noted as a serious issue. This is a national dilemma in South Africa and cannot be resolved in the short term by individual manufacturers.

6.7   CONCLUSION

According to the results, South African automotive industry is facing a huge number of challenges. However, majority of the challenges are largely outside the control of industry. From the aforementioned, developing strategies that will overcome these challenges is of high importance as competitiveness partially depends on a smooth supply chain management system, ranging from suppliers of raw materials, the transformation process, WIP, final product, and shipments to customers.

ACKNOWLEDGMENTS

The authors acknowledge the support provided by the University of Johannesburg, and the supervisors that facilitated the success of the study.

REFERENCES

Arnold, J. T., & Chapman, S. N. (2004). Introduction to Materials Management (5th edition). Chapel Hill: North Carolina State University, Pearson.

Bennett, D., & O'Kane, J. (2006). Achieving business excellence through synchronous supply in the automotive sector. *Benchmarking: An International Journal*, 13(1–2), 12–22.

Caddy, I. N., & Helou, M. M. (2007). Supply chains and their management: Application of general systems theory. *Journal of Retailing and Consumer Services*, 14(15), 319–327.

Choi, T., & Krause, D. (2006). The supply base and its complexity: Implications for transaction costs, risks, responsiveness, and innovation. *Journal of Operations Management*, 24, 637–652. https://doi.org/10.1016/j.jom.2005.07.002.

Council of Supply Chain Management Professional. (2018). Retrieved March 1, 2022, from: https://www.supplychainquarterly.com/articles/1665-the-evolution-of-cscmp.

Duarte, S., & Machado, V.C. (2011). Manufacturing paradigms in supply chain management. *International Journal of Management Science and Engineering*, 6(5), 328–342.

Fawcett, S. E., Ellram, L. M., & Ogden, J. A. (2007). *Supply Chain Management: From Vision to Implementation*. Upper Saddle River, NJ: Prentice Hall.

Gansler, C., Luby, R. E. Jr., & Kornberg, B. (2004). Supply chain management in government and business. In Gansler, J. & Luby, J. R. (eds) *Transforming Government*. Bethesda, MD: IBM Centre for the Business for Government Series.

Hugo, W. M. J., Badenhorst-Weiss, J. A., & Van Biljon, E. H. B. (2004). *Supply Chain Management: Logistics in Perspective*. Pretoria: Van Schaik.

Humphreys, P. K., Huang, G., Cadden, T., & Mcivor, R. (2007). Integrating design metrics within the early supplier selection process. *Journal of Purchasing & Supply Management*, 13, 42–52.

Humphrey, J., & Salerno, M. S. (2000). Globalisation and assembler-supplier relations: Brazil and India. In *Global Strategies and Local Realities* (pp. 149–175). London: Palgrave Macmillan.

Humphrey, J., & Schmitz, H. (2000). *Governance and Upgrading: Linking Industrial Cluster and Global Value Chain Research* (Vol. 120, pp. 139–170). Brighton: Institute of Development Studies University of Sussex.

Ikome, J., Ayodeji, S. P., & Kanakana, G. (2015). *The Effects of Disruption on Different Types of Tile Manufacturing Industry-Layouts: An Empirical Investigation on Tile Manufacturing Industry.* 2015 Portland International Conference on Management of Engineering and Technology (PICMET), Portland, pp. 1929–1936.

Ikome, J. M., & Kanakana, G. M. (2018). *A literature review of South African automotive industry global competitiveness.* Portland International Conference on Management of Engineering and Technology (PICMET), Portland, 1–4. doi: 10.23919/PICMET.2018. 8481934.

Ikome, J. M., & Laseinde, O. T. (2020). The global constrains of South African automotive industry and a way forward. In Markopoulos, E., Goonetilleke, R., Ho, A., & Luximon, Y. (eds) *Advances in Creativity, Innovation, Entrepreneurship and Communication of Design. AHFE 2020: Advances in Intelligent Systems and Computing* (vol. 1218). Cham: Springer. https://doi.org/10.1007/978-3-030-51626-0_27.

Ikome, J. M., Laseinde, T., & Katumba, M. G. (2021). An empirical review and implication of globalization to the South African automotive industry. In *International Conference on Applied Human Factors and Ergonomics* (pp. 263–270). Cham: Springer, July.

Ikome, J. M., Laseinde, O. T., & Katumba, M. G. K. (2022). The future of the automotive manufacturing industry in developing nations: A case study of its sustainability based on South Africa's paradigm. *Procedia Computer Science*, 200, 1165–1173.

Kaplinsky, R., Morris, M., & Readman. J. (2012). Understanding upgrading using value. *Chain Analysis*, 3.

Khayundi, F. (2011). Existing records and archival programs to the job market. *Journal of the South African Society of Archivists*, 44, 62–73.

Leenders, M. R., & Fearon, H. E. (2004). *Purchasing and Supply Chain Management* (11th edition). Chicago: Irwin.

Leenders, M. R., & Fearon, H. E. (2012). *Purchasing and Supply Chain Management* (13th edition). Chicago: Irwin.

Lockström, M., Schadel, J., Harrison, N., & Moser, R. (2009). *Status Quo of Supplier Integration in the Chinese Automotive Industry: A Descriptive Analysis.* Proceedings of 18th IPSERA Conference, 5–8 April, Wiesbaden, Germany, 1315–1327.

Lysons, K., & Gillingham, M. (2003). *Purchasing & Supply Chain Management* (6th edition). Harlow: Prentice Hall.

Morrison, A., Pietrobelli, C., & Rabellotti, R. (2008). Global value chains and technological capabilities: A framework to study industrial innovation in developing countries. *Oxford Development Studies*, 36(1), 39–58.

Naude, M. J., & Badenhorst-Weiss, J. A. (2011a). Supplier–customer relationships: Weaknesses in South African automotive supply chains. *Journal of Transport and Supply Chain Management*, 6(1), a33. https://doi.org/10.4102/jtscm.v6i1.33.

Naude, M. J., & Badenhorst-Weiss, J. A. (2011b). Supply chain management problems at South African automotive component manufacturers. *Southern African Business Review*, 15(1).

Pires, S., & Cardoza, G. (2007). A study of new supply chain management practices in the Brazilian and Spanish auto industries. *International Journal of Automotive Technology and Management*, 7(1), 72–87.

Sturgeon, T., & Van Biesebroeck, J. (2012). Related information globalisation of the automotive industry: Main features and trends. *International Journal of Technological Learning, Innovation and Development*, 4.

Swink, M., Melnyk, S. A., Cooper, M. B., & Hartley, J. L. (2011). *Managing Operations across the Supply Chain.* New York: McGraw-Hill.

Trade and Investment South Africa (TISA) and the Motor Industry Development Council (MIDC), (2003). *Current Developments in the Automotive Industry, 2003, Department of Trade and Industry*, 7th report, Pretoria.

UNCTAD, United Nation. (2010). *Conference on Trade and Annual Development Report.* Retrieved March 1, 2022, from: https://unctad.org/en/docs/tdr2010_en.pdf.

Updike, K. (2012). *Supply Chain Constraints Present a Three-Part Challenge to Automotive Suppliers.* www.industryweek.com/supply-chain/supply-chain-constraints-present-three-partchallenge-automotive-suppliers.

Wang, Z., Zhang, R., & Liu, B. (2021). Rebate strategy selection and channel coordination of competing two-echelon supply chains. *Complexity*, 1–20.

7 Assembly Line Optimization Applying a Construction Algorithm

Jennifer Mayorga-Paguay, Jorge Buele,
Angel Soria, and Manuel Ayala-Chauvin

CONTENTS

7.1 INTRODUCTION

The first prototypes of the turbocharger were tested in automobiles in the United States. However, its development was slow due to the posed barriers that the technology available during that time could not overcome. During World War II, its implementation was carried out in US combat aircrafts, which made it possible to better appreciate its importance in the military field (Alfano, 1986). Despite their importance in avionics, turbochargers are currently used in the development of ground transportation. Among the main ones are cars, buses, trucks, tractors, and other vehicles with similar applications. This equipment increases engine power and torque by increasing the air–fuel mixture. In general, it is composed of a turbine and a rotary air compressor, which are in opposite sides of the same axis (Ono and Ito, 2021). Turbochargers enabled the development of smaller but more powerful engines, due to the usage of the gases generated by the engine itself (Sivagnanasundaram, Spence and Early, 2013). This has generated further research in the field of the automotive industry, as can be seen in the review by Lee et al. (2017). Heuer et al. (2008) made an attempt to improve its aerodynamic performance, preventing high cycle fatigue (HCF) failures.

In Latin America, industries that manufacture this equipment have been developed, but they have only been in existence for a few years. Development processes and methodologies are deficient, in some cases empirical, which motivates the search for techniques to improve internal processes. The reduction of production and waste

DOI: 10.1201/9781003383444-7

costs requires a worker–machine–material balance (Lozada-Cepeda, Lara-Calle and Buele, 2021). This is obtained by performing a good design of the distribution of people, materials, and activities in the plant (Ali Naqvi et al., 2016). Systematic Layout Planning (SLP) is a technique used for the efficient use of resources and to organize the workplace as well as the teams. In the works of Benitez, Da Silveira, and Fogliatto (2019) and Zhu and Wang (2009), its application can be seen in the design of health and forest production facilities. In the review of Archibald (2017), the use of SLP for the design of classrooms and their interaction with educators can be seen in 49 studies. In Zhou et al. (2010), the workshop design of an industry that manufactures motor-cycles based on this technique is carried out, obtaining an 18% increase in production. Similarly, Khariwal, Kumar, and Bhandari (2020) use SLP in railway construction to improve the flow between workshops and reduce movement in the workshop.

In the context, this chapter describes the implementation of the SLP methodol-ogy. As a case study, there is a manufacturing industry that is responsible for the development of turbochargers for all types of vehicles for individual use. As a start-ing hypothesis, it is proposed that a better distribution of activities in the plant will reduce production times. Proportionately, this symbolizes increased productivity and competition in the market.

7.2 METHODOLOGY

7.2.1 INITIAL DIAGNOSIS

A small company from Latin American with a few years in the market of equipment importing, merchandizing, and repairs of the HT3B turbocharger model for light vehi-cles has been chosen. For a period of two months, onsite observation was carried out in the workshop to study the structure of a turbocharger and to define the parts that compose its structure. Currently, the distribution of the company consists of six areas; nevertheless, emphasis will be placed on the assembly area. This process was divided into three stages, with 42 operations that comprise it. The format of the exposed dia-grams was designed based on the process flow proposed by Palacios Acero (2016). The route that the parts move to reach the worktable is 13 m long, the route to move the axis toward the balancer is 18 m long, and on the way back the route is 9 m long. The tool board is located 16 m from the assembly site, while the assembled turbocharger is 9 m away from the packaging area. In total, there is a route of 56 m, with an elapsed time of 00:09:11 minutes.

Most of the processes are handled manually, so it is required to define an assembly line. In this way, the departments that require time and transportation investment are strategically located nearby. Since there is no available information in the com-pany, the starting point is to design the current layout of the workshop, as shown in Figure 7.1.

In the link https://drive.google.com/file/d/1S1yBvwBdDSzV0dWaZHUt2kSdEYr sW7Wn/view?usp=sharing, the number of operations is shown and the flow diagram during the assembly process is shown in Figure 7.2. Assembling a turbocharger takes 56 minutes and 33 seconds, without considering delays due to machinery failure, since there have been no continuous breakdowns.

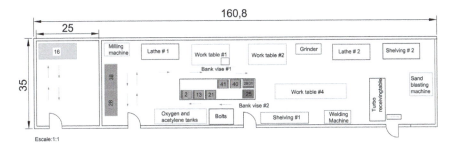

FIGURE 7.1 Initial workshop layer.

Source: Developed by the author.

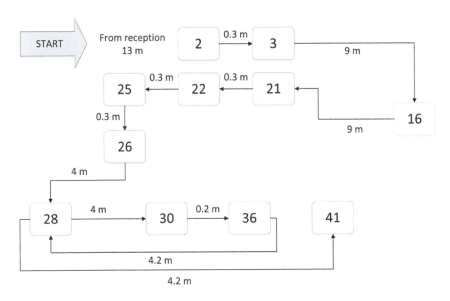

FIGURE 7.2 Flow diagram of the current assembly area.

Source: Developed by the author.

7.2.2 IMPROVEMENT PROPOSAL

The systematic layout problem (SLP) method was developed by Muther (1973) and consists of three phases: analysis, search, and solution. The first phase was already developed with the review and formulation of the problem: also reviewing the job distribution, flow of raw material, workers, etc. During the second phase, the three stages that were previously recognized are described, as well as the substages and departments that comprise them. Initially, there were 42 activities that were grouped according to their similarity in 19 departments. For example, department 1 includes rim, horn, collar, girdle, and counter-girdle placement operations.

Subsequently, departments are chosen in pairs and a label is established that reflects the degree of need to place them close to each other in the workshop. These labels are A, E, I, O, U, and X, which stand for absolutely necessary, especially important, important, indifferent, not important, and undesirable, respectively. A numerical value for these labels was provided as follows: A=6, E=5, I=4, O=3, U=2, and X=1. The sum of these results is in the total proximity ratio "T". The department that will be placed first is the one that has obtained a higher T value and in case of a tie it will be the one with the largest surface area. Tables 7.1, 7.2, and 7.3 show the data of stages 1, 2, and 3 respectively.

TABLE 7.1
Relationship Matrix of the Departments of the First Stage

Substage	No.	Department	T	1	2	3	4	5
Central body	1	Place rim, horn, collar, headband, against headband	14		A	O	O	U
	2	Put the crescent	17			A	O	U
	3	Place oil deflector	20				A	E
	4	Place seal and plaque	17					E
	5	Place the bottom bracket	14					

TABLE 7.2
Relationship Matrix of the Departments of the Second Stage

Substage	No.	Department	T	1	2	3	4	5	6
Axle and roll	1	Put rim on the axle	16		A	E	U	U	X
	2	Balance verification	17			A	O	X	X
	3	Axis insertion	19				O	O	U
Compression wheel	4	Install compression wheel	18					A	I
	5	Adjust nut	15						O
	6	Place elastic on the plate	11						

TABLE 7.3
Relationship Matrix of the Departments of the Third Stage

Substage	No.	Department	T	1	2	3	4	5	6	7	8
Intake casing	1	Place intake casing	26		A	I	O	O	U	A	U
	2	Attach admission headband	25			A	O	O	O	U	U
	3	Adjust intake nuts	25				O	O	U	U	E
Exhaust casing	4	Put exhaust casing	28					A	I	A	O
	5	Attach exhaust headband	28						E	O	E
	6	Adjust base nuts	24							A	U
Valve	7	Couple valve	31								A
	8	Adjust valve nuts	25								

7.3   RESULTS

Considering the geometry of each area and including the corridors and technical requirements, the last phase was carried out, where the possible solutions to the previous phase are evaluated. Based on the calculated T value, an approximation of the best location for each department was performed. However, to obtain results, it was decided to implement a construction algorithm. The CORELAP (COmputerized RElationship LAyout Planning) method, developed by Lee and Moore (1967), was selected. It is known as a pioneer among the various available software that enables an improvement in the distribution of the departments according to the proximity criteria that the designer provides. In this case, to optimize trajectories and easy-to-implement close relationships, the process was divided into three stages. If this division is not performed, the result is quite diffuse and almost impossible to make work in a real process. The graphic representation of the distribution of the departments of stage 1 is provided in Figure 7.3, stage 2 in Figure 7.4, and stage 3 in Figure 7.5.

Based on the distribution obtained by the CORELAP software, changes were applied to the analyzed process. First, five activities that are considered redundant have been eliminated, leaving 37 activities as a result, as shown in the operations diagram found at the following link: https://drive.google.com/file/d/1bJl mrv2iCU4h2ouaj0F38FC0g2Ep5fyf/view?usp=sharing. Based on this information, an assembly line that previously did not exist is structured and is described in Figure 7.6.

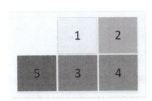

1. Place seal and plaque
2. Put the crescent
3. Place rim, horn, collar, headband, against headband
4. Place oil deflector
5. Place the bottom bracket

FIGURE 7.3   Internal department distribution from stage 1.

Source: Developed by the author.

1. Put rim on the axle
2. Check balance
3. Insert axis
4. Install compression wheel
5. Adjust nut
6. Place elastic on the plate

FIGURE 7.4   Internal department distribution from stage 2.

Source: Developed by the author.

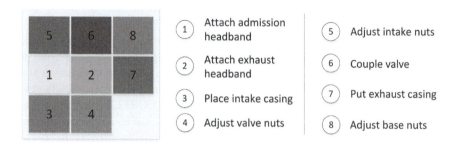

FIGURE 7.5 Internal department distribution from stage 3.

Source: Developed by the author.

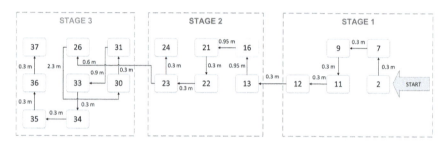

FIGURE 7.6 Flow diagram of the proposed assembly line.

Source: Developed by the author.

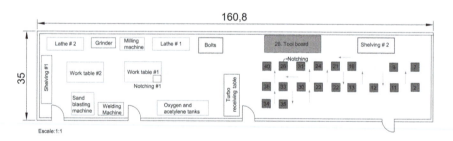

FIGURE 7.7 Proposed distribution layer for the workshop.

Source: Developed by the author.

In order to reduce the assembly line travel times, the three stages of the process are placed together consecutively and the displacement of the operator toward the balancing area is eliminated because the balancer will be located on the surface of stage 2. Meanwhile, the machinery that does not contribute to the process is displaced and placed in the remaining available surface. The worktables are divided, and their size is adapted according to the activity performed, taking into account the anthropomorphic measurements of the worker and their free movement. The layout of the current distribution of the proposed assembly area is presented in Figure 7.7.

7.4   CONCLUSIONS

From the starting point of this research proposal, the company did not have archived information on the activities that take place in the production of a turbocharger. Likewise, in terms of the distances that the staff travels in the assembly workshop and the times used in the fabrication process, there was no recorded information. As part of the applied SLP technique, the assembly diagram and data collection were performed to enable the identification of the operations that are required in the fabrication of the equipment. The operations involved are also analyzed and grouped into departments according to their similarities and redundancies: to ease the subsequent processes that the CORELAP software will carry out. In this direction, the relationships between departments are assessed and evaluated, to aid in the formulation of an efficient structural design for the workshop of this plant, which is embodied in a layout. After this intervention, there is a decrease in the staff travel routes and in the transportation to the warehouse and balancing areas.

The travel distance got reduced from 56 m to 16.2 m, thus, a reduction of 71.07% was achieved. Thus, for the case of time spent, the reduction recorded started from 9:11 minutes to 1:25 minutes, demonstrating a reduction of 84.58% from the previous value. Also, operations that had similarities were unified and new departments were developed. On the other hand, operations that are no longer necessary due to the new distribution were suppressed. As a result, the new assembly line has a production time spam of 28:37 minutes, which represents a reduction of the 49.4% of the original observed value resulting from the elimination of unnecessary trips. Workstations, machinery, and tool boards have been strategically changed to aid in the assembly process.

A balance between the variables of efficiency and cost optimization was defined. Based on the execution of this research study, the quality of the turbocharger parts will not be modified. Moreover, the efficiency of the production process will be increased. In preliminary studies, an increase of more than 30% was obtained. This is in line with the work of Zhou et al. (2010) where production increased by 18%. However, further tests are still required to be executed as future work. Thus, it is also proposed to use this methodology in other areas of the organization.

REFERENCES

Alfano, D.L. (1986) *Turbocharger applications*. SAE Technical Papers. doi:10.4271/862051.

Ali Naqvi, S.A. et al. (2016) 'Productivity improvement of a manufacturing facility using systematic layout planning', *Cogent Engineering*, 3(1). doi:10.1080/23311916.2016.1207296.

Archibald, L.M. (2017) 'SLP-educator classroom collaboration: A review to inform reason-based practice', *Autism & Developmental Language Impairments*, 2. doi:10.1177/2396941516680369.

Benitez, G.B., Da Silveira, G.J.C. and Fogliatto, F.S. (2019) 'Layout planning in healthcare facilities: A systematic review', *Health Environments Research and Design Journal*, 12(3), pp. 31–44. doi:10.1177/1937586719855336.

Heuer, T. et al. (2008) 'An analytical approach to support high cycle fatigue validation for turbocharger turbine stages', *Proceedings of the ASME Turbo Expo*, 1, pp. 723–732. doi:10.1115/GT2008-50764.

Khariwal, S., Kumar, P. and Bhandari, M. (2020) 'Layout improvement of railway workshop using systematic layout planning (SLP)-A case study', *Materials Today: Proceedings*, 44, pp. 4065–4071. doi:10.1016/j.matpr.2020.10.444.

Lee, R.C. and Moore, J.M. (1967) 'CORELAP—computerized relationship layout planning', *The Journal of Industrial Engineering*, 18(3), pp. 195–200.

Lee, W. et al. (2017) 'Overview of electric turbocharger and supercharger for downsized internal combustion engines', *IEEE Transactions on Transportation Electrification*, 3(1), pp. 36–47. doi:10.1109/TTE.2016.2620172.

Lozada-Cepeda, J.A., Lara-Calle, R. and Buele, J. (2021) 'Maintenance plan based on TPM for turbine recovery machinery', *Journal of Physics: Conference Series*, 1878(1). doi:10.1088/1742-6596/1878/1/012034.

Muther, R. (1973) *Systematic layout planning*. Cahners Books.

Ono, Y. and Ito, Y. (2021) 'Development of new generation MET turbocharger', *14th Proceedings of the International Conference on Turbochargers and Turbocharging*, pp. 242–251.

Palacios Acero, L.C. (2016) *Ingeniería de Métodos Movimientos y Tiempos*. Ecoe Ediciones.

Sivagnanasundaram, S., Spence, S. and Early, J. (2013) 'Map width enhancement technique for a turbocharger compressor', *Journal of Turbomachinery*, 136(6). doi:10.1115/1.4007895.

Zhou, K. et al. (2010) 'Study on workshop layout of a motorcycle company based on systematic layout planning (SLP)', *International Conference on Image Processing and Pattern Recognition in Industrial Engineering*, 7820, p. 78203R. doi:10.1117/12.867211.

Zhu, Y. and Wang, F. (2009) 'Study on the general plane of log yards based on systematic layout planning', *2009 International Conference on Information Management, Innovation Management and Industrial Engineering, ICIII 2009*, 3, pp. 92–95. doi:10.1109/ICIII.2009.332.

8 The Competitive and Productivity Challenges in Developing Nations
A Case Study of the South African Automotive Industry

John M. Ikome, Opeyeolu Timothy Laseinde, and M.G. Kanakana Katumba

CONTENTS

8.1 INTRODUCTION

During the early 1990s, the automotive market was gradually expanding to various continent and this provided myriad of opportunities for the reintroduction of South Africa (SA) as a nation to the global automotive market. However, South Africa's automotive industry finds itself affected strongly by a number of countries including Indian and China. Within SA, there are 120 first tier (TIER-1) automotive component manufacturing industries, of which 75% are multinational industries. Also, there are 200 second and third tiers (TIER-2 and TIER-3) locally owned industries [1]. Despite the large second- and third-tier numbers, the net value of local components used in locally assembled vehicles is less than 40% of the total component value [2].

Agreeing to Naude and Badenhorst-Weiss [3], automotive assemblers often import lower cost parts from overseas, and local automotive manufacturing firms in South Africa are not as competitive as manufacturers from India and China.

DOI: 10.1201/9781003383444-8

8.2 LITERATURE REVIEW

8.2.1 COMPETITIVENESS

India and China are advancing rapidly because of large inflows of foreign direct investment (FDI). Looking at their size and that of developing countries in general, it is evident that South Africa is a relatively small competitor in comparison. For this reason, it is imperative to analyze, identify, and develop possible competitive models or methods that can attempt to enhance South Africa's competitiveness and sustainable growth within the automotive industry.

Conferring to a research work done by Humphrey and Schmitz [4], all businesses are constantly under pressure to boost their efficiency and competitiveness. It's much more difficult for companies in developing countries to enter and succeed in the automotive industry, which is already a well-established, mature, and globalized industry.

Developing countries are competing against one another, especially China and India, whose upswing has serious implications for other developing countries. Given the sheer scale of these two countries and the immense emerging supply of high-skilled researchers, engineers, skilled workers, or labor force and technicians give them a greater competitive edge over developing countries. A research work done by Altenburg et al. [5] argue that China and India poses a great challenge to South Africa and also for the rest of the world due to their high-tech advancement. This implies that both Organization for Economic Co-operation and Development (OECD) and developing countries are threatened.

Furthermore, several developing countries are attempting to catch-up with this pace. Such countries include North Korea, Singapore, India, and Thailand that have done so successfully [6]. Small- and medium-sized enterprises (SMEs) from selected emerging countries have managed to build up competitive advantages, enabling them to compete productively in global markets and also understand what is needed to create competitive know-how is the aptitude to continuously upgrade their skills in order to increase their returns [7].

8.3 STATEMENT OF RESEARCH PROBLEM

In recent years, the South African manufacturing industries are dominated by transformation process as a transitional element for change since the inceptions of democracy in 1994. Globalization and the fourth industrial revolution has significantly influenced the growth of the automotive industry, including the appetite and changes in end users' demands in vehicle features [8–11]. Numerous studies have been conducted [12–14] that reveal the lack of competitive models in South African Automotive industry. As a result, these industries are faced with many problems in the areas of productivity output, supply chain, research, and development.

In addition, competitive strategies are the greatest part that defines the industries well-being within the constantly changing environment. In respect to this, many companies adopted innovation as a driving force to deal with those uncertainties, particularly the automotive; however, the lack of appropriate competitive methods

Automotive Producers Location

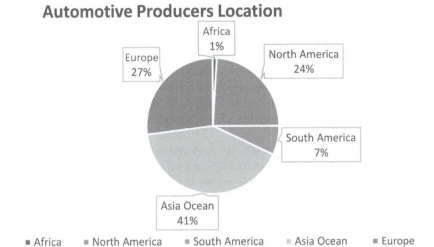

FIGURE 8.1   Geographical location of automotive producers.

Source: OICA 2017.

and strategies that can enhance a global competitiveness is a great challenge for the well-being of the organization and the nation as a whole.

Therefore, there is a need for a research in order to analyze and develop a competitive model that can enhance the automotive industries using the Fourth Industrial Revolution Imperatives.

According to Sturgeon and Van Biesebroeck [15], the automotive industry has grown globally from 65.4 million to 84.1 million of total new vehicles sold throughout 2005–2015 and much of the production has shifted to developing countries. Source from Organisation Internationale des Constructeurs d'Automobiles (OICA) [16] shows that production is now interspersed throughout the globe with facilities on every continent. This is illustrated in Figure 8.1 and it shows a clear indication of only 1% in Africa which is startling.

The automobile industry, both in the past and now, is regarded as a central industry, a peculiar economic anomaly that dominated the twentieth century. It has brought about previously unthinkable improvements in the way we live and function, and its goods continue to shape our culture and daily lives today. It is a cornerstone industry, a flag of economic growth, for most of the developed world, and increasingly for the developing world [17]. In the literature, the link between the global and national competitiveness of the automotive industries particularly in South Africa appears only for a few specific indicators of global competitiveness. This evidently shows the need for a comprehensive or holistic approach to explain the competitiveness in terms of all factors constituting South Africa's level of competitiveness in the automotive industry.

According to Barnes [3], South Africa automotive industry compares favorably with similar industry in developing countries with regard to availability to raw materials, flexibility in production, the support of government, and infrastructure. However, despite these positive aspects, the South African automotive industry's competitiveness is under severe pressure, as it faces challenges such as poor infrastructures, technology, and high labor costs.

This is further justified by a research work done by Naude and Badenhorst-Weiss [3] and Moodley, Morris, and Barnes [18], where they narrated the ordeal of South African Automotive Component Manufacturers (ACMs). They submitted that South African ACMs are competing against cheap imported parts, including counterfeit parts in some cases.

In addition, the total number of vehicles produced in Africa in 2014 was 586,396, of which 539,424 vehicles which equated to 92% of total vehicles were produced by South Africa automotive manufacturing industry for that year [16]. Although these figures appear convincing, in reality this is relatively small as it represents less than 1% in the international market share. Table 8.1 shows that a total number of 84,141 new vehicles were produced in 2014 and a 5.3% growth rate was realized. According to the statistic by OICA [16], during 2008–2014, the South African automotive shares increased steadily from 0.61% to 0.67%, but unfortunately declined rapidly to 0.07% in 2014.

The global automotive industry is directed by a global trend of development and satisfying the customers need or capacity demand; however, since the introduction of globalization, the automotive is experiencing huge challenges, which are induced by mergers and acquisition, of capacity, global manufacturing, strategies, outsourcing and insourcing, environmental requirement, innovation, and technology. As a result, all of these constraints have a major impact on emerging nations like the South Africa [20]. As a result of the aforementioned, the survival and growth of South African automotive industry, particularly (ACMs) global competitiveness, is vital for the growth of the nation.

According to a quarterly report from NAAMSA [21], the automotive industry's contribution toward global production decreased from 0.73% in 2010 to 0.60% in 2015. Therefore, the study seeks to understand the South African automotive industry competitiveness with the aim to identify various obstacles that prevent it from being competitive. Against the research background, the research problem of this study is read as the lack of proper competitive methods and strategies in South African

TABLE 8.1
Various Challenges within the South African Automotive Industry

Years	2010	2011	2012	2013	2014	2015	% Change
Global vehicle production (million)	72,150	70,760	61,710	77,610	79,990	84,141	+ 5,3%
SA vehicle production (million)	0,535	0,563	0,374	0,472	0,533	0,539	+ 1,3%
SA share of global production (%)	0,73%	0,8%	0,68%	0,61%	0,67%	0,6%	11,7%

Source: Organisaion Internationale des Constructuers d'Automobiles (OICA) [16, 19].

automotive industries, resulting in poor competitiveness and consequently affecting the country's economic well-being.

Even though numerous efforts have been made to improve the competitiveness of the South African automotive industry, much still need to be done because an optimal solution still need to be obtained. From the aforementioned, the key questions are as follows:

- What are the implications and barriers, preventing the South African automotive industry from being globally competitive?
- What is South African global competitiveness in the automotive industry?

8.3.1 RESEARCH OBJECTIVES

To sufficiently answer these questions, objectives were formulated as presented herein. The research aimed at the following:

1. Analyzing the South African automotive industry's global competitiveness and identifying the challenges and barriers that prohibit it from being globally competitive.
2. Investigating the global competitiveness of South Africa's automotive industry.

8.3.2 METHODOLOGY

Multifactor productivity (MFP) is an important industrial engineering tool which measures the productivity of all the resource (inputs) that are used within the manufacturing process, compared to the total output in units. Consequently, MFP gives a clear indication of the overall efficiency of the production process of an industry in terms of "economic" factors, and according to Jay Heizer and Barry Render [22], multifactor productivity can be calculated using Equation (8.1).

$$P = Q/M\_(a + M\_b + M\_a) = Q/M\_t \tag{8.1}$$

Here P = productivity function, Q = output quantity, and $I = \Sigma m\_t$, which is the input quantity and should be interpreted as valid for a system that is affected by many production factors as well as systems affected by single factors.

The input and output serve as constraint to production. From Equation (8.1), output quantity can be predicted using Equation (8.2):

$$PM\_t = Q \tag{8.2}$$

By differentiating Equation (8.2) using product role, we obtained how the input and output varies during production:

$$dQ = M\_t \; dp/dt + P \; dmt/dt \tag{8.3}$$

By redesigning Equation (8.3) with time interval of production of both sides of the expression, the rate of change becomes Equation (8.4):

$$dQ = M\_t\,(t) + P(t) \qquad\qquad (8.4)$$

The second differential equation obtained in Equation (8.3) is used to obtain random differential Equation (8.4). The "first integral" solution is the main expression that indicates the evolution of output per given time due to disruptions. This solution (i.e., output per given time) is actually the productivity. Thus, during practical application, the issue is on determining the fluctuation of the variables "M" and "P" found in the preceding expressions.

Negative variable input values are not considered in the present results and additionally, the increase in input (i.e., increase in input with time) during the manufacturing process is also not considered. Similarly, the situations that generate negative values of frequency of downtime are not considered. The various constraints that are used are as follows: a = 0.017, b = 10−6, and t = 1.06. These parameters are ideally chosen for trend indication and are actually limited on different operational needs. The results obtained from this analysis, including a well-structured questionnaire and interviews that were administered in a number of automotive industries in South Africa, are presented and discussed in the following section.

8.4   RESULTS AND ANALYSIS

By using engineering equation solver and the data obtained from the questionnaire and interviews, the results and findings of Equations (8.1)–(8.4) are discussed subsequently, followed by respective discussions per graphical depiction in Figure 8.2.

It can be observed in Figure 8.2a that as disruptions occur continuously with time, the total productivity decreases as time goes on. Figure 8.2b indicates that due to disruptions, the actual production time increases.

Using the approach of line balancing to determine the number of workstations (though not at the level of avoiding bottlenecks), the result of Figure 8.2c is obtained. This plot indicates that as time goes on and following continuous disruptions, the number of facilities/resources required should increase.

This increase in facilities/resources will account for the time lost due to disruptions so as to increase production rate. The Scheduler should reschedule following the trend that is only predicted by the nature of disruption. Figure 8.2d indicates that the relationship between the amounts of resources that should be rescheduled does not depend linearly on the actual production time.

The analysis of Tables 8.2 and 8.3 shows an instability of the total production (outputs) and sales of diesel and petrol cars, including medium commercial vehicles and electric and hybrid vehicles from 2015 to 2019 of the South African automotive industry. From the aforementioned, the results clearly show that there is a constraint with the global competitiveness of South African Automotive industry.

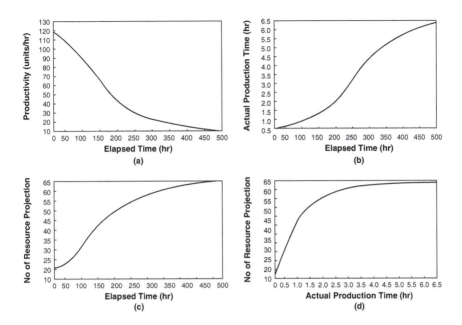

FIGURE 8.2   Time evolution of the (a) productivity, (b) production time, (c) number of work-station required, and (d) number of resources to meet demand.

TABLE 8.2
Petrol versus Diesel Passenger Cars and Light Commercial Vehicle Production and Sales during 2015–2019

Year	2015	2016	2017	2018	2019
Diesel cars and diesel light commercials	189,168	173,952	184,147	188,865	177,959
Petrol cars and petrol light commercials	397,430	345,829	346,918	335,705	330,204
Total cars and light commercials	586,598	519,781	531,065	524,570	408,163
Diesel vehicles as % of total	32,2%	33,5%	34,7%	36,0%	35,0%

TABLE 8.3
Diesel and Petrol Hybrid and Electric Cars Productivity and Sales during 2015–2019

	2015	2016	2017	2018	2019
Diesel cars	69,761	61,856	65,547	62,618	55,563
Diesel hybrid	0	0	0	0	0
Petrol cars	342,136	298,818	302,192	302,427	299,408
Petrol hybrid	502	574	306	144	253
Electric	79	41	68	58	154
Diesel light commercials	119.407	112,096	118,600	126,247	122,396
Diesel light commercials	55,294	47,011	44,726	33,278	30,796

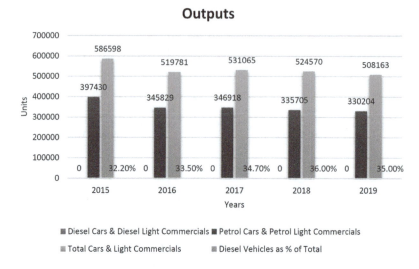

FIGURE 8.3   Productivity, outputs, and sales of cars and light commercial vehicles.

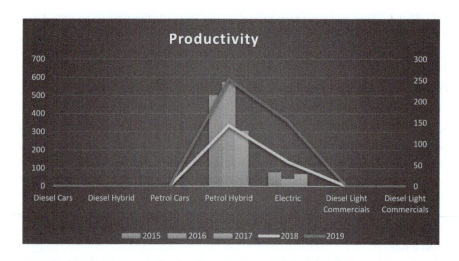

FIGURE 8.4   Hybrid and electric car output in South African automotive industry.

Furthermore, Figure 8.3 also elaborates clearly that productivity fluctuate as time goes on, especially between 2018 and 2019, and it is evident that within five years, the total cars and light commercial vehicles output in 2019 was 508,163 vehicles compared to 586,598 outputs in 2015. This shows a productivity decrease of 78,435 cars and commercial vehicles between 2015 and 2019.

The results presented in Figure 8.4 clearly show that hybrid and electric cars output or productivity is still very limited as this is a new technology, particularly the diesel hybrid which has a zero productivity.

8.5 CONCLUSION

The automotive industry of South Africa has grown to become a leading manufacturing industry in terms of economic growth within the domestic market. However, this industry faces huge amount of pressure related to competitiveness, as it has to face major automotive industries globally.

As a result, any improvement to its productivity and competitive strategy is very vital, including automotive components manufacturer (ACMs), as this will also improve employment in the automotive industry and economic growth. Additionally, the research findings reveal that South African automotive industry still need to improve in its productivity in order to gain competitive advantage.

ACKNOWLEDGMENT

We would first like to acknowledge the University of Johannesburg for the internal funds provided through the GES funding mechanism.

REFERENCES

(1) National Association of Automotive Component and Allied Manufacturers (NAACAM). (2018). *NAACAM Directory 2018*. Gauteng: NAACAM.

(2) NAAMSA. (2017). *NAAMSA Annual Report 2007*. Pretoria: National Association of Automobile Manufacturers of South Africa [Online]. Available at: http://naamsa.co.za/papers/ (Accessed: 18 May 2021).

(3) Naude, M. J., & Badenhorst-Weiss, J. A. (2012). Supplier-customer relationships: Weaknesses in South African automotive supply chains. *Journal of Transport and Supply Chain Management*, 6(1), 91–106.

(4) Humphrey, J., & Schmitz, H. (2016). *Governance and Upgrading: Linking Industrial Cluster and Global Value Chain Research*. Brighton: Institute of Development Studies.

(5) Altenburg, T. (2006). Governance patterns in value chains and their development impact. *The European Journal of Development Research*, 18(4), 498–521.

(6) Luo, J. (2015). *The Growth of Independent Chinese Automotive Companies*. Working Paper. Cambridge, MA: MIT International Motor Vehicle Program.

(7) Ferdowsi, M. A. (2010). UNCTAD–United Nations conference on trade and development. In *A Concise Encyclopedia of the United Nations* (pp. 698–705). Leiden: Brill.

(8) Metu, A. G., Madichie, C. V., Kalu, C. U., & Nzeribe, G. E. (2020). The fourth industrial revolution and employment in Sub-Saharan Africa: The role of education. *Journal of African Development*, 21(1), 116–137.

(9) Ikome, J. M., Laseinde, O. T., & Kanakana Katumba, M. G. (2021). *An Empirical Review and Implication of Globalization to the South African Automotive Industry*. 12th International Conference on Applied Human Factors and Ergonomics and the Affiliated Conferences, San Francisco.

(10) Radziwill, N. M. (Reviewed). (2018). The fourth industrial revolution. *Quality Management Journal*, 25(2), 108–109. doi: 10.1080/10686967.2018.

(11) Ikome, J. M., & Kanakana, G. M. (2018). *A Literature Review of South African Automotive Industry Global Competitiveness*. 2018 Portland International Conference on Management of Engineering and Technology (PICMET), Honolulu, 1–4.

(12) Department of Trade and Industry (DTI). (2012). *Creation of Rebate Governing the Automotive Production and Development Programme*. Report No. 419. Pretoria: The International Trade Administration of South Africa.

(13) Ikome, J. M., & Laseinde, O. T. (2020). *The Global Constraints of South African Automotive Industry and a Way Forward.* 11th International Conference on Applied Human Factors and Ergonomic (AHFE), San Francisco.

(14) Mutsiya, M., Steyn, J., & Sommerville, J. (2008). *Concurrent Engineering and the Automotive Supplier Industry in South Africa.* PICMET 2008 Proceedings, Cape Town, South Africa, 1265–1272.

(15) Sturgeon, T., Memedovic, O., Van Biesebroeck, J., & Gereffi, G (2009). Globalisation of the automotive industry: Main features and trends. *International Journal Technological Learning, Innovation and Development*, 2(1), 7–24.

(16) Organisaion Internationale des Constructuers d'Automobiles (OICA). (2018). *Statistic.* International Organization of Automotive Manufacturer. https://www.oica.net/statistics/.

(17) Maxton, G. P., & Wormald, J. (2004). *Time for a Model Change: Re-Engineering the Global Automotive Industry.* Cambridge: Cambridge University Press, ISBN 0521837154, 9780521837156.

(18) Barnes, J., & Morris, M. (2018). Staying alive in the global automotive industry: What can developing economies learn from South Africa about linking into global automotive value chains? *European Journal of Development Research*, 20(1), 31–55.

(19) Organisaion Internationale des Constructuers d'Automobiles (OICA). 2017. *Operation Management*, 11th edition. *Statistic.* International Organization of Automotive Manufacturer. https://www.oica.net/statistics.

(20) Automotive Industry Export Council (AIEC). (2016). *Automotive Export Manual 2016—South Africa Pretoria.* www.aiec.co.za.

(21) National Association of Automotive Component and Allied Manufacturers (NAACAM). (2012). *The Authority of the South African Automotive Components Industry.* Sandton: NAACAM.

(22) Heizer, J., & Render, B. (2014). *Operations Management*, eleventh edition. New York: Pearson Education.

9 Digital Twin, Servitization, Circular Economy, and Lean Manufacturing
A Step toward Sustainability

Junaid Ahmed

CONTENTS

DOI: 10.1201/9781003383444-9

9.1 INTRODUCTION

Among sustainable development goals (SDGs), the most important and significant is SDG No. 09 Industry, Innovation, and Infrastructure. The SDG No. 09 directly addresses to industrial practices of manufacturing and subsequent generation of industrial waste and pollution. It is argued that current production management and control system, which is defined as a system of converting flow of inputs such as information and material to produce output in the form of product and services to be consumed by end user, is responsible for such waste and pollution (Kumar & Suresh, 2006). It is also said that majority of environmental problems being faced today by humanity is the result of waste and pollution caused by industries. Thus, tackling such industrial waste and pollution is necessary globally. The SDG No. 09 completely addresses to current problems of industry, its waste generation practices, and many others, specifically cost-effectiveness. The SDG No. 09 provides a clear direction to address such problem by changing production management and control system through innovation and building resilient infrastructure of converting input into outputs (Pradhan et al., 2017; Sinha et al., 2020).

The production management and control system as defined earlier is referred to as planning and controlling of industrial process such as flow of material, information, and finances in an effective way so that a value is to be created for the customers in shape-needed product and services. Zhuang et al. (2018) have illustrated a historical development of production management and control system. According to them, production management and control system is divided into four different stages. The first stage started in the early 1960s in which single point management with each function of production was being managed separately. The second stage which started early in the 1990s developed concept of integration among different activities of production. The third stage started in the 2000s that evolved concept of production management by bringing collaboration in between firms. Finally, fourth stage also referred as smart and intelligent manufacturing defined as automation of flow of material and information to create a product and services which provide maximum value to consumers. In the literature, smart and intelligent manufacturing is mentioned into wider context of Industry 4.0. The latest innovation of smart and intelligent manufacturing innovation as contemporary production management and control system is being described as critical to achievement of SDG No. 09 (Mittal et al., 2019).

The Industry 4.0 is defined as "industrial infrastructure which tends to embed different technologies, such as machine, electrical, operational and information technology, with the purpose to create efficient, effective and sustainable products and services" (Mrugalska & Ahmed, 2021, p. 10). The industrial infrastructure of Industry 4.0 consists of certain key technologies that form the basis of new industrial infrastructure. It is widely concluded in the literature that Industry 4.0 has played a key role in the development of production management and control system critical for the achievement of SDG No. 09. Although production management and control system such as lean manufacturing which originated at Toyota automobile company has promised to achieve objectives of SDG No. 09, although lean manufacturing's wide range adoptability seems to be a problem that has to be addressed. However, increasing amount of literature points out that lean manufacturing can now be possible under Industry 4.0 industrial infrastructures and smart and intelligent manufacturing. Apart from lean manufacturing, other production management and control systems such as servitization and circular economy are also pointed out by experts as possible way to achieve objective enshrined in SDG no. 09 (Dantas et al., 2021).

The technologies which are part of Industry 4.0 infrastructure are of cutting-edge nature. It includes cloud computing, virtual and augmented reality, automated guided vehicle, sensors, robots, Internet of Things (IoT), big data and analytics, simulation machine-to-machine communication, cyber physical systems, and artificial intelligence. Each of the technologies has different application in a wide range of fields. The manufacturing, especially smart and intelligent manufacturing, is also possible due to integration of various Industry 4.0 technologies. One key technology which is highly necessary for smart and intelligent manufacturing is referred to as digital twin. The digital twin is not a new technology, apart from those mentioned earlier. However, digital twin is an innovative integration of some of Industry 4.0 technologies such as cyber-physical system, simulation, big data and analytics, cloud computing, and virtual and augmented reality, etc. The literature argues that digital twin is not just necessary for development of smart and intelligent manufacturing, but it also plays an important role in developing and implementing production management and control systems which promise sustainability as enshrined in the SDG No. 09 (Sepasgozar, 2021; Horváthová et al., 2019).

The purpose of this chapter is to provide an account from existing literature on Industry 4.0 technology of digital twin as prerequisite for developing smart and intelligent manufacturing. Further, smart and intelligent manufacturing is all about digitalization and automation of value chain activities, but it can also offer implication for sustainability of manufacturing. The various production management and control system exits in the literature that promises sustainability in manufacturing. The present research theorizes that technology such as digital twin can be corroborated with the production management and control system that supports manufacturing sustainability as enshrined in SDG No. 09. This chapter in its theorization provides a detailed account of three different, innovative, and highly effective production management and control systems that support sustainability. This includes circular economy, servitization, and lean manufacturing. This chapter theorizes and summarizes that digital twin technology which develops smart and intelligent manufacturing also helps to create and develop circular economy, servitization, and lean manufacturing necessary for achieving SDG No. 09. The following sections illustrate further detail and account on this discussion.

9.2   LITERATURE REVIEW

9.2.1   Digital Twin

In the recent years and particularly mainstreaming of Industry 4.0 concept, interest in the digital twin gas increased in both academic and industry communities. The widely agreed and followed definition of digital twin is one that is given by the CIRP Encyclopedia of Production Engineering. They have described digital twin as follows:

A digital twin is a digital representation of an active unique product (real device, object, machine, service, or intangible asset) or unique product-service system (a system consisting of a product and a related service) that comprises its selected characteristics, properties, conditions, and behaviors by means of models, information, and data within a single or even across multiple life cycle phases.

(Sami et al., 2019)

The digital twin is described as key technological infrastructure which is highly necessary for implementing smart and intelligent manufacturing (Jones et al., 2020). Liu et al. (2021) have developed comprehensive list of technologies which is needed in order to develop architecture of digital twin (Vijayakumar, 2020).

The concept of digital twin can be traced back to the 1960s in NASA Apollo project. The NSAS created two identical spaceships referred to as "twin". One spaceship was left to the space and another was kept here on Earth. Spaceship that was kept here at Earth used to mirror spaceship which was performing the mission. After that, Michael Grieves in his class of product life cycle management in 2003 at University of Michigan introduced the concept of a "virtual, digital representation equivalent to a physical product" (Grieves, 2005). The central concept of Grieves product life cycle management was that information on physical entity such as product can be mirrored for better level of analysis. Although Grieves model was not referred originally to as digital twin but it contained all aspects of it. Thus, it can be said that Grieves argument can be treated as conceptual building block and origin of digital twin concept (Grieves & Vickers, 2017). Zhuang et al. (2018, p. 1153) argue on conceptual development of digital twin that it "is a virtual, dynamic model in the virtual world that is fully consistent with its corresponding physical entity in the real world and can simulate its physical counterpart's characteristics, behavior, life, and performance in a timely fashion".

The digital twin has huge implication in production management and control system such as smart and intelligent manufacturing and others. He and Bai (2021) concluded in their excellent review study that digital twin in smart and intelligent manufacturing incorporate Industry 4.0 technologies such as cyber-physical system, big data analytics, IoT, cloud computing, 3D design, and sensors. These technologies and their integration as digital twin have been further employed for the various purposes in developing smart and intelligent manufacturing production management and control system. It includes product design, production process simulation, product manufacturing, assembly line, equipment status monitoring, product fault warning and maintenance, etc. It can be noted that activities highlighted by He and Bai (2021) in which digital twin implicates in production management and control are in majority. Similarly, Ma et al. (2020) have also concluded in their review research that digital twin–based production management system helps to create production management by incorporating and streamlining activities such as product design, product manufacturing, and intelligent services management. Thus, overlapping research conclude that digital twin is an essential integration of technologies necessary in developing smart and intelligent manufacturing (Zhuang et al., 2018). Liu et al. (2021) have also given detailed account on use of digital twin in product life cycle and production management phases.

9.2.2 SERVITIZATION

The phenomena of servitization are not new or more specifically emerged along with Industry 4.0. The term was initially coined by Vandermerwe and Rada (1988). In defining servitization, they argued that

> Since the primary objectives of business are to create wealth by creating value; "servitization" of business is very much a top management issue. It gets firms beyond "servicing" where a good is repaired or maintained by the manufacturer or "moments of truth" where the firm comes face-to face with its customer. It involves a different strategic thrust, level of organizational complexity and an order where the old traditional managerial recipes no longer fit.

> (Vandermerwe & Rada, 1988, p. 315)

Thus, servitization was defined as "offering fuller market packages or 'bundles' of customer-focused combinations of goods, services, support, self-service, and knowledge" (Vandermerwe & Rada, 1988, p. 314).

The servitization is becoming highly important in today's ultracompetitive industrial environment. Servitization enables value offering differentiation, deterring competitors from imitating firm's value proposition, and enhancing customer satisfaction and loyalty (Martin et al., 2019). The servitization has also enabled firms in their own strategic planning and decision-making way. With the help of Industry 4.0 technologies such as digital twin, servitization has helped the firm to offer "integrated product-service offerings are distinctive, long-lived, and easier to defend from competition based in lower cost economies" (Baines et al., 2009). Thus, it can be argued that bundles of values which are offering integrating products with services can help a firm to become highly competitive in the market and allows it to stay relevant in the longer run (Kamal et al., 2020). Further, it can also be noted that servitization does bring about certain challenges such as bringing agility in operation and production management (Mrugalska & Ahmed, 2021), implementing product-service philosophy, and providing integrated value offerings and internal processes and capabilities (Martinez et al., 2017).

The servitization has increasingly been referred to as one of most important production management and control system. The servitization offers new perspectives to firm in designing, manufacturing, and delivering the values (Lightfoot et al., 2013). This new perspective requires firms to channelize resources and inputs such as material, information, and finance in quite a different way. The servitization which solely focuses upon providing superior value to customer needed to redesign entire operation management architecture (Kowalkowski et al., 2017). Baines et al. (2009) have provided detailed guidelines on production management and control aspect of servitization. They have concluded that to implement servitization as production management system, certain structural characteristics needed to be adjusted and incorporated which includes process and technology, operation management capacity, redesigning of manufacturing facilities, and redesigning of planning and control. Apart from changes in structural characteristics, it is needed to bring adjustment in the infrastructural characteristics which includes human resources, products and services range, performance measurement, customer relations, and supplier relations.

9.2.3 CIRCULAR ECONOMY

The concept of circular economy has attracted wide range of interests among researchers of sustainability, management sciences, and policymakers. It is being argued for the circular economy that it presents a unique opportunity to solve problem of sustainability and provides source of next-generation creative destruction innovation. Since the conception of circular economy as concepts, various definition and conceptual understanding have come up. Kirchherr et al. (2017) have reviewed more than 114 definitions of circular economy. So, it can be argued that circular economy is being defined by different researchers and on different paradigms and perspectives. However, Geissdoerfer et al. (2017) have come up with an innovative definition of circular economy. According to them, circular economy is defined as "a regenerative system in which resource input and waste, emission, and energy leakage are minimized by slowing, closing, and narrowing material and energy loops. This can be achieved through long-lasting design, maintenance, repair, reuse, remanufacturing, refurbishing, and recycling" (Geissdoerfer et al., 2017, p. 3). From definition, it is established that circular economy is based upon the concept of loop economy and supply chain which is meant to reduce the waste generated by the industrial economy. It has been argued that circular economy can result in benefits such as waste reduction, local job creation, resource efficacy, and dematerialization of the manufacturing economy (Geissdoerfer et al., 2017). It is being argued about the circular economy that it is one of the sought sustainability business models with potential to solve waste and emission problem.

By looking at conceptual underpinning of the circular economy, it can be argued that circular economy is one of the innovative and promising sustainability-oriented business models or production management and control systems. The circular economy from production management perspective enables manufacturing and remanufacturing of existing products. More specifically, it helps to design products which are based upon raw material of existing products being used and disposed by consumers. Such practices not just address problem of waste but also promise to utilize resources highly efficiently. On the other hand, another concept emerged in the circular economy literature which refers to creation of by-product and co-product. The co-product is referred to as separate product, but it can easily be produced using the same manufacturing and operation management process (Lewandowski, 2016). The co-product helps in many ways by saving energy and materials and eliminating new waste and supply chain efficiency. To realize this, it needed to focus more upon designs of original and co-products. It has been further argued that circular economy is highly efficient in saving and conserving energy at aggregate level. To realize the motto of circular economy, an innovation is needed at both business model level and production management aspects. Such innovation incorporates a clear goal of circular economy model. It needed to incorporate the specific purpose which can be remanufacturing co-product, by-product, or recycling of existing products and design production management system accordingly (Bocken et al., 2016).

9.2.4   Lean Manufacturing

The concept of lean and its implication in manufacturing is a revolutionary one. The concept of lean is associated with and originated at Toyota, which manufactures automobiles in a highly innovative way which enhanced its competitiveness globally. The concept in academic literature was popularized through a book titled *The Machine that Changed the World* (Womack et al., 2007). The concept of lean manufacturing is based upon the philosophy of lean. The lean philosophy of production is based on five set of principles which include defining customer value, defining value stream, making it "flow", establishing pull, and finally striving for excellence (Womack et al., 2007). Thus, based upon the principle of lean philosophy, lean manufacturing can be defined as follows:

> Lean production is a philosophy of production that emphasizes the minimization of the amount of all the resources (including time) used in the various activities in the enterprise. It involves identifying and eliminating non-value adding activities in design, production, supply-chain management, and dealing with the customers. The lean manufacturing strives to shorten the time line between the customer order and the shipment of the final product, by consistent elimination of waste.
>
> (Cox & Blackstone, 1998)

The lean manufacturing and philosophy is also considered as one of the promising production management and control systems that not just helps industries enhance cost-efficiency, maximize value, and create differentiation, but also promises sustainability. The lean manufacturing implementation has been one of the biggest challenges which is set to be resolved with the help of Industry 4.0 technologies such as digital twin. According to Mrugalska and Wyrwicka (2017, p. 471):

> Lean production successfully challenged the mass production practices to the production systems focused on good quality products aimed at customers' satisfaction, where everything that does not add value is concerned to be waste. It can be the answer to a great flexibility of production systems and processes realizing complex products and supply chains. In order to achieve it, it is advisable to introduce IT integration of the production level with the planning level, customers and suppliers by CPS known as Industry 4.0.

The lean manufacturing also has an implication for the sustainable and responsible manufacturing. The lean philosophy is inherently consistent with sustainability purpose. Lean manufacturing espoused to address the sustainability problem through efficient use of resources, energy conservation, and reduce waste (León & Calvo-Amodio, 2017).

9.3   DIGITAL TWIN AND SERVITIZATION

The servitization is becoming a popular business model innovation in the range of industries from automobile to health sectors. In fact, it is inferred from literature that servitization is becoming key mix of strategy to pursue organization's longer term mission and goals. However, realizing such strategy, i.e., implementing servitization as tool of value offering, has been a key problem for organization. The literature suggests introducing operation management structure and work culture that supports servitization has been a key hurdle for organization. Further, it was required by organization to rethink their value chain and enable the process that serves customer in a better way. All of these challenges were significant in nature and it required investment which poses challenges to organizations. However, with innovation of Industry 4.0 and its enabling technologies such as Digital Twin, implementing servitization as production management system and value chain is more than easy. The digital twin has enabled organization to design structure and deliver their value offering using servitization. It can be argued that digital twin is becoming an important enabler to servitization. It must be noted that literature is highly limited on relationship of Digital Twin and Servitization. Thus, systematic review studies are gap in the literature needed to be filled (Woitsch et al., 2022).

From scant literature, it has been observed that digital twin and technologies such as cyber-physical system, Industrial Internet of Things (IIoT), sensor, big data analytics, cloud computing, and artificial intelligence can endow or completely transmute the aspects of services being delivered. The product-based value offering can be completely transformed through product services system (PSS). Such adjustment in value offering can result in innovation of new business model and revenue streams. The servitization enabled by the digital twin on the other hand has also contributed significantly to enhancing consumer experiences of product/services under the ambit of PSS. The consumers are being delivered with the value which was not envisioned previously using integration of both servitization and digital twin. A good example would include the platform-based PSS which has completely transformed the value associated with particular products and services in a different and efficient way (West et al., 2020).

Emphasis is needed to be put upon manufacturing or physical products-based services. Every manufacturer of physical product, in order to enhance their competitiveness and creation perception superior value of their offering in minds of consumer, is adopting services and more specially servitization. The servitization enables firms to protect from intense competition and a term which in marketing literature is referred to as Blue Ocean. The Blue Ocean market refers to the area of competition where only few seller or manufacturer compete for large chunk of customers. The few in that market is possible due to high value offering of which servitization is an important part. Further, a new term Everything-as-a-Service (XaaS) as services is being mentioned in the literature which is referred to as delivery of everything as value to consumer in the form of services. Based upon the concept of XaaS, integration of digital twin with operation can spur physical products into services based upon ideas of on-demand use of physical products and returning to original owner (Company) on need satisfaction, dynamic reconfiguration, and platform independence (Meierhofer et al., 2020; Leng et al., 2021).

9.3.1 Application of Digital in Servitization Value Creation

Despite benefits and potential of servitization enabled through digital twin technologies, insight is still limited on application of digital twin in creating value. The literature searched up till date has not informed on companies using and integrating digital twin with servitization concept to create value for the customers. Thus, it presents a research gap which is necessary to be filled. This chapter attempts to highlight an insight based upon limited literature on different ways in which digital twin can be integrated with servitization in creation of better value offering. Bertoni and Bertoni (2022) provided insight on way digital twin technologies can be applied in conjunction with servitization. They have categorized into various stages of product/ services life cycle.

9.3.1.1 Ideation

The first stage in the product life cycle is developing an idea. Bertoni and Bertoni (2022) have provided an insight into integration of digital twin with servitization in idea generation. They have provided three different ways in which integration of digital twin with servitization is applied in idea generation. First, physical to virtual (P2V) and virtual to physical (V2P). Some of best ideas for servitization-based value offering can be developed by imagining all of products which are being developed or manufactured physically by machines to be converted into virtual such as 3D manufacturing, etc. In the same way, all of the products/services which are being developed or manufactured virtually can be offered to consumer physically using technology such as augmented and virtual reality (A&VR). According to Bertoni and Bertoni (2022, p. 3):

> The combination of the P2V and V2P connections allows for a continuous optimization cycle, as possible physical states are predicted in the virtual environment and optimized for a specific goal. That is, once a virtual optimization process determines an optimal set of virtual parameters, these are propagated through to the physical twin. In turn, the latter responds to the change, and the loop cycle continues. The frequency by which this process repeats is indicated as "twinning rate", and in an ideal world, this shall be near instantaneous.

Second, new servitization-based product can be imagined using digital technology by leveraging upon the customer data. It is argued that digital twin enables twining by converting and generating data from physical products and services and experience of consumer with the same into virtual space. Such data can be stored in virtual space with the help of technology such as big data and analytics and cloud computing. Thus, by analyzing such data and plotting on the simulative technologies, a range of new servitization-based value offering can be imagined. Finally, new servitization-based products can also be imagined by twining data from physical entity (PE) and virtual entity (VE). The physical entity can be various such as product itself, environment into which product is consumed, and life cycle of product. The data can be replicated into virtual entity and many ideas can be generated for new products and services (Niu & Qin, 2021).

9.3.1.2 Design

In the product life cycle, the second stage after the product creation comes design. Digital twin enables to design the products based upon servitization philosophy. The ideas to design such as digital mockups and digital replicas of physical products and services can be very effective. It helps designer to understand a very important feature which can provide greater value and experience (Peruzzini et al., 2016; Bertoni & Bertoni, 2022). Zhang et al. (2019) have identified four different steps to convert ideas into designed product and services. They have concluded that four steps include, first, development of platform for new product; second, data acquisition and preprocessing; third, data analytics for service and product innovation; and finally, digital twin–enabled service innovation. Further, as digital twin is an open access platform, consumer can be enabling to make an addition into already developed design or create a new design.

9.3.1.3   Realization, Deployment, and Support

The automated manufacturing is one of the examples of servitized and digitally enabled realization of mass production of designed products. The technologies such as 3D manufacturing, social, and other forms of manufacturing are prime examples. The cyber-physical system along with 3D and additive manufacturing technologies enables a digitally designed product to be produced physically and it contain precise measurement with which a product is to be produced and delivered to consumers (Bertoni & Bertoni, 2022; Olivotti et al., 2019; Bonnard et al., 2019).

9.3.1.4   Over Product Life Cycle

The digital twin also provides a solution in using product and services over its life cycle. The digital twin can be described as enabler in providing maintenance and repair of products and services in hand. The digital twin technology also helps consumer in maintaining the product effectively. From data collected on the usage of product, it can be apprised to consumer about the current health of a product, whether it is working healthy or it is deteriorating; and if it is deteriorating, what is needed to restore the product and services. The continuous generation of data from product usage can help consumer in various ways with regard to effective product handling and increasing overall life of a product (Abramovici et al., 2018; Schuh et al., 2018).

9.4   DIGITAL TWIN AND CIRCULAR ECONOMY: A ROLE OF INFORMATION HANDLING AND MANAGEMENT

To fight problems such as waste and sustainability of resources for humanity, circular economy model is being positioned as one of solution. The importance of circular economy as business model has already been summarized. In this section, an attempt is being made to analyze role of digital twin and enabling technologies in spurring process of circular economy. Although very limited scientific literature is available on the subject, it is concluded here that more research are needed in highlighting the role of digital twin in enabling circular economy process and business models.

The circular economy business model operates on philosophy of regeneration of existing resources, minimizing waste, leakages of energy, inputs and materials, and developing a product which are long-lasting enough (Geissdoerfer et al., 2017). The data plays a very critical role on each of the operational aspects of circular economy which includes energy usage and leakages, inputs and materials of products, and design of product to make it long-lasting. The data would also be needed on various important aspects such as regenerative process in which new products are being created out of current material through both refurbishing and remanufacturing of products from not new component and raw materials. As pointed out by Preut et al. (2021, p. 1) on circular economy for the purpose of regenerative system of any product and other aspect:

> Depending on the specific product, it is often impossible to predict when the use phase of a product will end due to a defect or a user decision and if or when a product will be sent for reprocessing, reuse or recycling.

Thus, it can be argued that, information on potential circularity of product is very necessary to implement objective of circular economy and digital twin can be helpful in implementing such objectives.

Preut et al. (2021, p. 1) have concluded on the role of various actors in the circular economy: "As a link between individual phases and actors, logistics and material flow management are essential for the implementation of material and product cycles" in circular economy. They have concluded from the previous research some ten various actors who play an active role in the circular economy's regenerative system. It includes product manufacturer, provider/distributor, user, sharing system provider, maintenance/repair service provider, reuser, refurbisher/remanufacturer, recycler, logistics service provider, and governmental and regulatory actors. These actors of circular economy have to work in harmony among them by providing on time and accurate information to other members of circular economy. Such information management will be highly critical for undertaking circular activities. The traditional information management system has proven to be effective in handling information sharing among actors but digital twin can enhance entire process of information generation, sharing, and management among actors (Bressanelli et al., 2022).

As described by Preut et al. (2021, p. 10):

> [T]he digital twin is a virtual collection of information regarding a specific product and its entire lifecycle—from the design phase to end-of-life management, e.g., in terms of recycling. It enables visualization of design parameters and status data of a real product based on its data model. In case of changes in the state of the real product, its virtual image can be automatically updated through near-real-time transmission and evaluation of sensor data. The digital twin thus enables direct and vivid monitoring of its real counterpart. Information can be exchanged across different processes and actors and throughout the entire life cycle. Processes and product condition become more transparent and traceable.

It was further argued that "The stakeholders described above can add relevant information to this information collection and extract the information they need for the successful and efficient implementation of their respective processes" (Preut et al., 2021).

9.4.1 PERSPECTIVE FRAMEWORK OF DIGITAL TWIN AND CIRCULAR ECONOMY

The digital twin can be helpful for the circular economy at various stages of product life cycle such as design, production and remanufacturing, distribution, consumption, repairing, and reusage (Stahel, 2016). The digital twin can be helpful in two different ways to circular economy. First, it will be helpful in generating, handling, and disseminating information among the participants so that they can undertake necessary actions for implementing circularity in their respect. This process of digital twin handling information has been discussed in the preceding section very briefly. In this section, a layman framework is being introduced with purpose of twining various activities of circular economy and their perspective benefits. The following section provides details.

9.4.1.1 Design

A lot of energy in circular economy business model is being positioned upon design of product. The design must consider various elements such as material requirement and their sourcing using circulatory principles. The business needs to design their product in circular economy in a way that they must be able to procure components, material, and inputs easily using circularity as principle. The digital twin technologies can be very helpful in designing a product and assembling all of the components, material, and inputs circularity. The twin of real product design at virtual space makes it easier for designer to collect and analyze the data. The insight through twining will lead to make adjustment and changes deemed necessary for product use, reuse, and manufacturing (Awan & Sroufe, 2022).

9.4.1.2 Production and Remanufacturing

The production and continuous remanufacturing of products on the basis of circular philosophy is an important part of operation management. The digital twin technology by creating a virtual twin of physical product can help producer to assess the relative quality and functionality of product before commercialization. The data collected will help to asses' commercial and functional potential of circular products and services. The digital twin will also help to undertake changes which are deemed necessary in the manufacturing and delivery of product during phase of commercialization. The digital twin finally will also assess to manage the supply in an appropriate way for the current and ongoing commercialized product. Thus, it can be argued that, digital technologies enable circular organizations to assess potential of products manufacturing and remanufacturing (Reslan et al., 2022).

9.4.1.3 Supply and Distribution

The distribution of circular product is an important task to be accomplished. The digital twin technologies will enable organization to develop supply chain activities which is not just effective in terms of achieving market equilibrium but it will also help to save a lot of cost to be associated with supply chain. The digital twin will first help organization to create a virtual supply chain network of physical channel. The data collected and analyzed will help to optimize supply chain and distribution network further (Reslan et al., 2022).

9.4.1.4 Consumptions

The consumption of an action undertaken by consumer determines the firm's profitability and success of any new innovation. The prediction of consumer needs, wants, preferences, and satisfaction are important aspects of overall consumption action by consumers. The prediction of consumer needs, wants, preferences, and satisfaction can be made

> more effective by enforcing Digital twin to track the customer behavior dynamically such as the products they consume, their satisfaction. So instead of relying on the historical data, the data for digital twin will be from CRMs, logs, order processing info etc. Right product at right time can be achieved by creating suitable machine learning models on this dynamic dataset and this trained model are held in the digital twin, which runs them in real time.
>
> (Vijayakumar, 2020, p. 1)

9.4.1.5 Repair and Reuse

The companies in circular economy require the information as a user of a new product consumes the product. It would also be ideal to include information about use traces, flaws, and the resultant function limits, as well as safety risks. When deciding whether to buy a used item, at the very least, that knowledge is relevant. The information will also be required on the availability of take-back mechanisms at the end of the usage phase, much like the user of a new product, to keep the product in cycle. The digital twin technologies help companies to smoothen such process of data collection, information, and product handling which makes use and reuse of product easy and effective (Llorente-González & Vence, 2020).

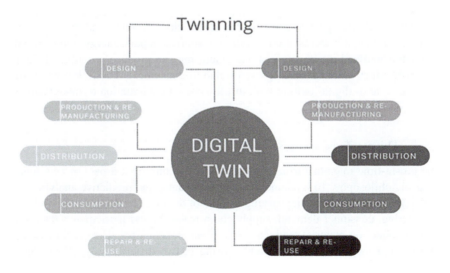

FIGURE 9.1 Circular economy and digital twin.

9.5 DIGITAL TWIN AND LEAN MANUFACTURING

The lean manufacturing compared to circular economy and servitization is also considered one of the most effective production management system as far as sustainability objectives are concerned. The lean manufacturing practices established in Toyota company post–Second World War gave Toyota an edge of competitiveness. The lean manufacturing is also touted as very necessary tools to achieve various sustainable development goals (SDG) such as SDG No. 09 and SDG No. 12.3. The research on the lean manufacturing has taken a pace since the inception of concept. Many innovation, technologies, and tools have been referred as enabling techniques of lean manufacturing. However, Industry 4.0 and enabling technologies such as cyber-physical system, Industrial Internet of Things (IIoT), sensor, simulation, and automated guided vehicles have been instrumental for organizations to develop automated and intelligent manufacturing which is essential for implementing lean philosophy in manufacturing. Among the very important, digital twin as literature review concluded is among best integration of Industry 4.0 technologies. The digital twin can help implement the lean manufacturing very effectively and smoothly. The literature on intersection of digital twin and lean manufacturing is highly limited and it is needed that more research has to be undertaken in this regard (Tran et al., 2021; Hartini & Ciptomulyono, 2015).

The limited literature on digital twin and lean manufacturing has summarized that the two most important functions of lean manufacturing is referred to as just-in-time manufacturing and automated manufacturing which is sometimes referred to as the literature as lean automation. The just-in-time manufacturing has enabled firm to produce goods and services which are only demanded by consumers and remove all those activities which are not adding value to production process often referred to as waste. The automation of manufacturing often intelligent manufacturing is considered as second most important aspect of lean manufacturing. The automated and intelligent lean manufacturing develops products and services which are highly efficient in terms of utilization of resources, inputs, energy, and materials. The automated and intelligent lean manufacturing can produce products which are very cost-effective and provide greater value to consumers. The following section discusses role of digital twin in both just-in-time and automation of manufacturing (Pawlewski et al., 2021).

9.5.1 DIGITAL TWIN AND JUST-IN-TIMES

The just-in-time manufacturing and supply are an important aspect of lean manufacturing philosophy. To implement just-in-time, data, and predictive analytics will play huge amount of role. The varying amount of data will be needed to be collected on ordering, consumer demand, supply chain network, and participant's data, data from suppliers on material, and data from other stakeholders. The digital twin can help to implement just-in-time manufacturing in two different ways. First, digital twin is composed of data-related technologies such as cloud computing, big data, and analytics. These technologies in digital twin architecture will enable a company to collect the data and analyze effectively and generate an insight which can feed production management system to achieve just-in-time objectives. Second, digital twin by creating an entire virtual supply chain and product of physical supply chain and product can help companies to track relevant development and analyze bottlenecks that can hamper future delivery of products and services on the basis of just-in-time. The twining of supply chain and product can also identify wastes virtually using simulation technologies which can easily be eliminated from physical supply chain and products manufacturing system (Tran et al., 2021).

9.5.2 AUTOMATED LEAN MANUFACTURING AND DIGITAL TWIN

The automated and intelligent manufacturing system is becoming core principle to apply for lean manufacturing philosophy. The automated and intelligent manufacturing produce product by optimally using component, energy, and materials. The new technology of additive and 3D manufacturing has further enhanced application of lean philosophy in the manufacturing. The digital twin is becoming key form of technology in automating the manufacturing system often on the basis of lean philosophy. As argued by He and Bai (2021, p. 5):

> The intelligent manufacturing technology is the deep integration and integration of the information technology, intelligent technology, and equipment manufacturing technology. The intelligent manufacturing technology is based on advanced technologies, such as the modern sensing technology, network technology, automation technology, and anthropomorphic intelligence technology. The intelligent manufacturing technology can realize intelligent design process, intelligent manufacturing process, and intelligent manufacturing equipment through intelligent sensing, human-computer interaction, decision making, and execution technology.

All of these technologies which are necessary for intelligent and automated manufacturing are working more effectively under the architecture of the digital twin. The digital twin enabled intelligent and automated manufacturing comprises assembly line and shop floor, necessary equipment, information system, and related service. The digital twin enabled organization to collect data from manufacturing platform and operation usage of equipment on that platform with help of key digital twin technologies such as cloud computing, artificial intelligence, Industrial Internet of Things, etc. The digital twin broadly collects the data related to manufacturing wider ecosystems such environmental, economic, and social data and integrate it with consumer, shop floor, and technology data. The digital twin also based upon the data from various sources develops virtual and physical prototyping map with automated and intelligent manufacturing equipment, shop floor, information systems, and manufacturing services (Leng et al., 2020).

9.6 CONCLUSION

The sustainable manufacturing of products and services is a goal which can be achieved once appropriate technologies and production management and control system are implemented in the organizations. This chapter summarized the existing literature on key Industry 4.0 technologies of digital twin and its role in establishing various manufacturing and production control system that can promise the sustainability, including servitization, circular economy, and lean philosophy of production and management. The literature does support role of servitization, circular economy, and lean in creating sustainable production system. However, literature is especially empirical and still limited on role of digital twin in enhancing these manufacturing and production control systems. Thus, a more research of empirical nature is needed to be developed which studies the role of digital with regard to servitization, circular economy, and lean manufacturing.

The literature has concluded about the servitization that it not just only facilitate sustainability in terms of production but also consumption as servitization-based production management system can reduce overall waste drastically. The servitization is based upon product services system (PSS) which is basically a framework to servitize every product into services. For example, instead of buying your own air-condition (AC) as asset in summer season, a consumer can get AC as services for time period in which he/she will be using it, i.e., summer season and once need for AC is over, it can be returned to original owner, i.e., company. The rent/price for AC would be highly economical given the charges of maintaining, repairing, and refurbishing for consumers. At aggregate level, it can drastically reduce over waste generated from disposal of AC products and few people are actually buying it and disposing it. The company which will have higher resources and knowledge can refurbish AC easily. The digital twin as literature shows it will be very helpful in the entire process of servitization based upon philosophy of PSS. First, digital twin using technology of simulation, sensor, cloud computing, and big data analytics can help company to generate an idea for product to be servitized. The virtual demonstration using simulation and feeding data and resultant analytical insight can provide clear strategies on how a particular product can be servitized. It will provide clear demonstration of consumer contact with product as services and journey of consumption. Second, digital twin can help company to create a twin of its servitized products. The twining will be helpful in observing changes in the product which has been servitized to customers using technology of sensor and cyber-physical system. The company, on the other hand, can undertake various actions to deliver smooth services to the consumers. Therefore, it can be said that digital twin can play significant role in enabling servitization-based production management system.

The circular economy has been touted as one significant innovation in production management system that promises sustainability. The circular economy helps to realize sustainable production and consumption of products through idea of regeneration of prevailing resources, minimizing waste and leakages of energy, inputs, and materials, and developing a product which are long-lasting enough. The digital twin technologies can be very helpful in enabling such regenerative process. The literature has argued that circular economy data and information management play an active and significant role. In regenerative process of circular economy, a necessary role has to be played by various actors in subsequent manner. The two problems are very important to be assessed here. First, design of actors, their activities, and their relationship among themselves. Second, information sharing and management and actions of actors based upon such information. The literature reviewed here argues that digital twin can be very helpful in enabling circular economy's actors activities based upon information management. The digital twin first through various technologies can help company to create network of actors which is highly efficient in handling their activities and, second, by twining the company's actual network of actors and their activities virtually, an effective monitoring and management system can be developed which help company to take a quick decision during situation of anomaly and current circular economy system can be upgraded. Further, literature review has also summarized that digital twin can be very helpful in deigning, producing, distributing, consumption, and reusing product based upon idea of regenerative system and long-lasting. A framework has been proposed which is presented in Section 9.4.1.

The lean manufacturing can also be a very important system of production management which can help us to achieve the goal of sustainable production and consumption. The lean manufacturing posits to eliminate every activity, material, and input, which does not add any value to product and consumption experiences. Such kind of activities of elimination of non-value activities, material, and input had led use resources efficiently and sustainably. The literature argues that achieving such lean manufacturing depends upon two concepts, i.e., just-in-time and lean automaton. The digital twin helps companies in both just-in-time and automation. The just-in-time is production management concept under ambit of lean manufacturing which makes no use of any inventory or less inventory. The complete process of just-in-time manufacturing depends upon the effective utilization of information. Digital twin can be very helpful in managing just-in-time supplies and distribution using technologies of sensor, cyber-physical system, cloud computing, and big data analytics. The digital through these technologies can create virtual representation of just-in-time activities and monitor these activities virtually. Thus, a real-time monitoring of just-in-time can be observed and it can be further enhanced. The digital twin can also be an important enabler of lean automation. The literature reviews have concluded that one of the first and significant applications of digital twin has been in enabling both automated and intelligent manufacturing as technologies necessary for creating automated and intelligent manufacturing shop floor such as cyber-physical system, simulation sensor, cloud computing, big data and analytics, IIOT, and automated guided vehicle works better way under the ambit of digital twin. The digital twin automates activities from conception to disposal of product by consumers.

Finally, it is concluded that digital twin is very important amalgamation of Industry 4.0 technologies and harnessing it can be helpful in enabling sustainable production and consumption. The servitization, circular economy, and lean manufacturing have been referred to as production management systems which can offer important implication for both sustainable production and consumption. The digital twin facilitates implementation of these important production management systems. The digital twin simplifies most important activities in each of production management systems such as deigning, producing, delivering, consumption products, information management, communication between actors and stakeholders, and automation. Each of these production management systems addresses to the problem of sustainability in a different and innovative way and digital twin helps in enabling their activities.

152
Human Factors in Engineering

REFERENCES

Abramovici, M., Savarino, P., Göbel, J. C., Adwernat, S., & Gebus, P. (2018). Systematization of virtual product twin models in the context of smart product reconfiguration during the product use phase. *Procedia CIRP*, *69*, 734–739.

Awan, U., & Sroufe, R. (2022). Sustainability in the circular economy: Insights and dynamics of designing circular business models. *Applied Sciences*, *12*(3), 1521.

Baines, T. S., Lightfoot, H. W., Benedettini, O., & Kay, J. M. (2009). The servitization of manufacturing: A review of literature and reflection on future challenges. *Journal of Manufacturing Technology Management*, *20*(5).

Bertoni, M., & Bertoni, A. (2022). Designing solutions with the product-service systems digital twin: What is now and what is next?. *Computers in Industry*, *138*, 103629.

Bocken, N. M., De Pauw, I., Bakker, C., & Van Der Grinten, B. (2016). Product design and business model strategies for a circular economy. *Journal of Industrial and Production Engineering*, *33*(5), 308–320.

Bonnard, R., Hascoët, J. Y., & Mognol, P. (2019). Data model for additive manufacturing digital thread: State of the art and perspectives. *International Journal of Computer Integrated Manufacturing*, *32*(12), 1170–1191.

Bressanelli, G., Adrodegari, F., Pigosso, D. C., & Parida, V. (2022). Circular economy in the digital age. *Sustainability*, *14*(9), 5565.

Cox III, J. F., & Blackstone Jr, J. H. (1998). *APICS Dictionary* (9th edition). Falls Church.

Dantas, T. E., De-Souza, E. D., Destro, I. R., Hammes, G., Rodriguez, C. M. T., & Soares, S. R. (2021). How the combination of circular economy and industry 4.0 can contribute towards achieving the sustainable development goals. *Sustainable Production and Consumption*, *26*, 213–227.

Geissdoerfer, M., Savaget, P., Bocken, N. M., & Hultink, E. J. (2017). The circular economy–a new sustainability paradigm? *Journal of Cleaner Production*, *143*, 757–768.

Grieves, M. W. (2005). Product lifecycle management: The new paradigm for enterprises. *International Journal of Product Development*, *2*(1–2), 71–84.

Grieves, M. W., & Vickers, J. (2017). Digital twin: Mitigating unpredictable, undesirable emergent behavior in complex systems. In *Transdisciplinary perspectives on complex systems* (pp. 85–113). Springer.

Hartini, S., & Ciptomulyono, U. (2015). The relationship between lean and sustainable manufacturing on performance: Literature review. *Procedia Manufacturing*, *4*, 38–45.

He, B., & Bai, K. J. (2021). Digital twin-based sustainable intelligent manufacturing: A review. *Advances in Manufacturing*, *9*(1), 1–21.

Horváthová, M., Lacko, R., & Hajduová, Z. (2019). Using industry 4.0 concept–digital twin–to improve the efficiency of leather cutting in automotive industry. *Quality Innovation Prosperity*, *23*(2), 1–12.

Jones, D., Snider, C., Nassehi, A., Yon, J., & Hicks, B. (2020). Characterising the digital twin: A systematic literature review. *CIRP Journal of Manufacturing Science and Technology*, *29*, 36–52.

Kamal, M. M., Sivarajah, U., Bigdeli, A. Z., Missi, F., & Koliousis, Y. (2020). Servitization implementation in the manufacturing organisations: Classification of strategies, definitions, benefits and challenges. *International Journal of Information Management*, *55*, 102206.

Kirchherr, J., Reike, D., & Hekkert, M. (2017). Conceptualizing the circular economy: An analysis of 114 definitions. *Resources, Conservation and Recycling*, *127*, 221–232.

Kowalkowski, C., Gebauer, H., Kamp, B., & Parry, G. (2017). Servitization and deservitization: Overview, concepts, and definitions. *Industrial Marketing Management*, *60*, 4–10.

Kumar, S. A., & Suresh, N. (2006). *Production and operations management*. New Age International.

Leng, J., Liu, Q., Ye, S., Jing, J., Wang, Y., Zhang, C., Zhang, D., & Chen, X. (2020). Digital twin-driven rapid reconfiguration of the automated manufacturing system via an open architecture model. *Robotics and Computer-Integrated Manufacturing, 63,* 101895.

Leng, J., Wang, D., Shen, W., Li, X., Liu, Q., & Chen, X. (2021). Digital twins-based smart manufacturing system design in industry 4.0: A review. *Journal of Manufacturing Systems, 60,* 119–137.

León, H. C. M., & Calvo-Amodio, J. (2017). Towards lean for sustainability: Understanding the interrelationships between lean and sustainability from a systems thinking perspective. *Journal of Cleaner Production, 142,* 4384–4402.

Lewandowski, M. (2016). Designing the business models for circular economy—towards the conceptual framework. *Sustainability, 8*(1), 43.

Lightfoot, H., Baines, T., & Smart, P. (2013). The servitization of manufacturing: A systematic literature review of interdependent trends. *International Journal of Operations & Production Management, 33.*

Liu, M., Fang, S., Dong, H., & Xu, C. (2021). Review of digital twin about concepts, technologies, and industrial applications. *Journal of Manufacturing Systems, 58,* 346–361.

Llorente-González, L. J., & Vence, X. (2020). How labor-intensive is the circular economy? A policy-orientated structural analysis of the repair, reuse and recycling activities in the European Union. *Resources, Conservation and Recycling, 162,* 105033.

Ma, J., Chen, H., Zhang, Y., Guo, H., Ren, Y., Mo, R., & Liu, L. (2020). A digital twin-driven production management system for production workshop. *The International Journal of Advanced Manufacturing Technology, 110*(5), 1385–1397.

Martin, P. C. G., Schroeder, A., & Bigdeli, A. Z. (2019). The value architecture of servitization: Expanding the research scope. *Journal of Business Research, 104,* 438–449.

Martinez, V., Neely, A., Velu, C., Leinster-Evans, S., & Bisessar, D. (2017). Exploring the journey to services. *International Journal of Production Economics, 192,* 66–80.

Meierhofer, J., West, S., Rapaccini, M., & Barbieri, C. (2020, February). The digital twin as a service enabler: From the service ecosystem to the simulation model. In *International Conference on Exploring Services Science* (pp. 347–359). Springer.

Mittal, S., Khan, M. A., Romero, D., & Wuest, T. (2019). Smart manufacturing: Characteristics, technologies and enabling factors. *Proceedings of the Institution of Mechanical Engineers, Part B: Journal of Engineering Manufacture, 233*(5), 1342–1361.

Mrugalska, B., & Ahmed, J. (2021). Organizational agility in industry 4.0: A systematic literature review. *Sustainability, 13*(15), 8272.

Mrugalska, B., & Wyrwicka, M. K. (2017). Towards lean production in industry 4.0. *Procedia Engineering, 182,* 466–473.

Niu, X., & Qin, S. (2021). Integrating crowd-/service-sourcing into digital twin for advanced manufacturing service innovation. *Advanced Engineering Informatics, 50,* 101422.

Olivotti, D., Dreyer, S., Lebek, B., & Breitner, M. H. (2019). Creating the foundation for digital twins in the manufacturing industry: An integrated installed base management system. *Information Systems and e-Business Management, 17*(1), 89–116.

Pawlewski, P., Kosacka-Olejnik, M., & Werner-Lewandowska, K. (2021). Digital twin lean intralogistics: Research implications. *Applied Sciences, 11*(4), 1495.

Peruzzini, M., Mengoni, M., & Raponi, D. (2016, August). How to use virtual prototyping to design product-service systems. In *2016 12th IEEE/ASME international conference on mechatronic and embedded systems and applications (MESA)* (pp. 1–6). IEEE.

Pradhan, P., Costa, L., Rybski, D., Lucht, W., & Kropp, J. P. (2017). A systematic study of sustainable development goal (SDG) interactions. *Earth's Future, 5*(11), 1169–1179.

Preut, A., Kopka, J. P., & Clausen, U. (2021). Digital twins for the circular economy. *Sustainability, 13*(18), 10467.

Reslan, M., Last, N., Mathur, N., Morris, K. C., & Ferrero, V. (2022). Circular economy: A product life cycle perspective on engineering and manufacturing practices. *Procedia CIRP*, *105*, 851–858.

Sami, C., Luc, L., Gunther, R., & Tolio, T. A. M. (2019). *CIRP encyclopedia of production engineering*. Springer.

Schuh, G., Jussen, P., & Harland, T. (2018). The digital shadow of services: A reference model for comprehensive data collection in MRO services of machine manufacturers. *Procedia CIRP*, *73*, 271–277.

Sepasgozar, S. M. (2021). Differentiating digital twin from digital shadow: Elucidating a paradigm shift to expedite a smart, sustainable built environment. *Buildings*, *11*(4), 151.

Sinha, A., Sengupta, T., & Alvarado, R. (2020). Interplay between technological innovation and environmental quality: Formulating the SDG policies for next 11 economies. *Journal of Cleaner Production*, *242*, 118549.

Stahel, W. R. (2016). The circular economy. *Nature*, *531*(7595), 435–438.

Tran, T. A., Ruppert, T., Eigner, G., & Abonyi, J. (2021, May). Real-time locating system and digital twin in Lean 4.0. In *2021 IEEE 15th international symposium on applied computational intelligence and informatics (SACI)* (pp. 000369–000374). IEEE.

Vandermerwe, S., & Rada, J. (1988). Servitization of business: adding value by adding services. *European Management Journal*, *6*(4), 314–324.

Vijayakumar, D. S. (2020). Digital twin in consumer choice modeling. In *Advances in computers* (Vol. 117, No. 1, pp. 265–284). Elsevier.

West, S., Meierhofer, J., Stoll, O., & Schweiger, L. (2020). *Value propositions enabled by digital twins in the context of servitization* (pp. 152–160). Advanced Services for Sustainability and Growth, Aston University.

Woitsch, R., Sumereder, A., & Falcioni, D. (2022). Model-based data integration along the product & service life cycle supported by digital twinning. *Computers in Industry*, *140*, 103648.

Womack, J. P., Jones, D. T., & Roos, D. (2007). *The machine that changed the world: The story of lean production-Toyota's secret weapon in the global car wars that is now revolutionizing world industry*. Simon and Schuster.

Zhang, H., Ma, L., Sun, J., Lin, H., & Thürer, M. (2019). Digital twin in services and industrial product service systems: Review and analysis. *Procedia CIRP*, *83*, 57–60.

Zhuang, C., Liu, J., & Xiong, H. (2018). Digital twin-based smart production management and control framework for the complex product assembly shop-floor. *The International Journal of Advanced Manufacturing Technology*, *96*(1), 1149–1163.

10 Reference Low-code Development Platform Architecture

Aurea BPM

Robert Waszkowski

CONTENTS

10.1 INTRODUCTION

Figure 10.1 shows the overall low-code platform architecture. The architecture assumes the division of the platform into subsystems organized in the development layer, as well as in the system management and integration layers (Sanchis, 2019).

The development layer of the platform consists of the design and development, metadata repository, deployment, and runtime subsystems. The most important of these are design, development, and runtime (Woo, 2020). The first allows the development of models in the form of diagrams and metadata, while the second one is responsible for launching the application in the end-user environment.

Models developed in the design and development subsystem are stored in the metadata repository subsystem, which is additionally responsible for providing controlled user access to model and metadata resources (Braude, Bernstein, 2011).

DOI: 10.1201/9781003383444-10

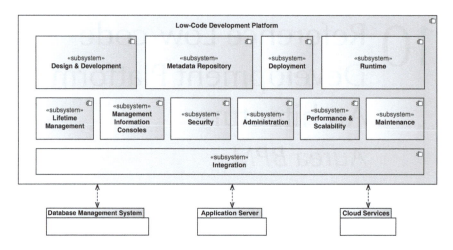

FIGURE 10.1   Overall low-code platform architecture.

Source: Own elaboration.

The developed and validated models are prepared and turned into executable form in the deployment subsystem. The runtime subsystem uses such prepared models to deliver the ready-to-use application to end users.

The system management layer consists of the following subsystems: lifetime management, management information consoles, security, administration, performance and scalability, and maintenance. In low-code development platform, the system management layer supports both development and runtime modules. This is not a typical approach and low-code platforms have to deal with management separately on these two different levels.

The integration layer is responsible for building and running interfaces with various systems in the same IT environment. When building an application supporting the company's business processes, ensuring cooperation with the systems within an enterprise is a very important element of implementation. It is estimated that the average business process cooperates in a more or less automatic way with several to a dozen IT systems. Managing this collaboration is a task of the integration layer.

This chapter is an extended version of the article published in the AHFE 2022 Conference Proceedings Edited Books (44 Volume Set) in the series of Applied Human Factors and Ergonomics International (ISSN 2771–0718).

10.2   DESIGN AND DEVELOPMENT SUBSYSTEM

Figure 10.2 presents all modules of the design and development subsystem. These modules can be divided into two groups: mandatory and optional. The required modules are BPMN process modeler, process data modeler, business rules, role management, UI designer. Optional modules that occur only in some low-code development platforms (McKendrick, 2017) are visibility matrix, multi-faceted modeler or mathematical model integrator.

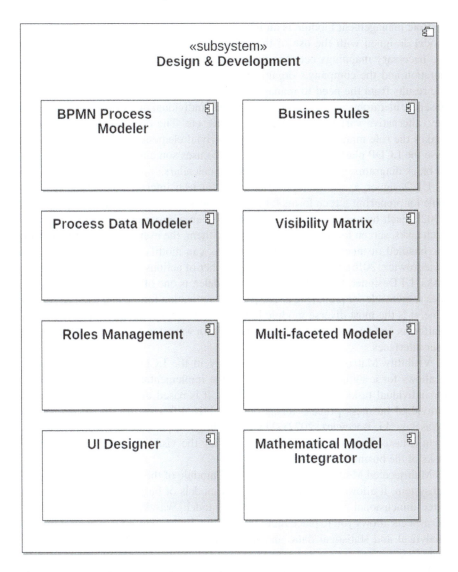

FIGURE 10.2 Design and development subsystem.

BPMN Process Modeler is a required component for all LCDPs designed to build applications based on business process models (Waszkowski, 2018). It is used for developing and storing business process models. Process Data Modeler is used to model the data used by the application to support business processes (Nowicki, 2016).

Business Rules define aspects of a business activity by introducing certain constraints into the system, usually resolved as true or false (Waszkowski, Kiedrowicz, 2015). They are used to ensure the correctness of business structures or to control or influence business behavior.

Role management module is an important element that complements the process model designed with the use of the process modeling notation. It allows defining the necessary mappings between the pools and swimlanes available in the BPMN notation and the company's organizational structure. The complexity of the module results from the need to manage changes of task contractors resulting from the absence, hierarchy, and order of substitutions, individual and group powers of attorney, alternative variants of acceptance paths, etc. The complexity of the problem makes the role management module a non-trivial element of the IT system, in this case the LCDP platform. This subsystem also uses some modeling techniques such as block diagrams, state diagrams, and team calendars.

UI Designer, as the module responsible for user interface modeling, is responsible for preparing screen forms for handling business process tasks (Guerrero, Lula, 2002). Each human task must be handled by an appropriate employee. To handle such tasks, screen forms are used. Thanks to them, the user has access to the data of the handled instance of the business process, can modify this data and documents (Kiedrowicz, 2016), and has access to a number of actions related to the handling of tasks. UI Designer, like BPMN Process Modeler, is one of the most important modules responsible for preparing a ready-made application for the end user. At the same time, it is the most diverse module in terms of its implementation in various LCDP platforms. In fact, each manufacturer has its own way of modeling and implementing user interfaces.

Visibility Matrix is a module found only in the LCDP Aurea BPM platform. It allows for a significant acceleration of the implementation of the user interface for individual tasks of the business process. It is based on the assumption that the individual business process forms only slightly differ in the set of handled data (Waszkowski, Bocewicz, 2022). Thanks to this, it is possible to construct one basic form (Master Form) and parameterize it using the visibility matrix for subsequent tasks of the business process.

Multifaceted Modeler is another optional module of the Design and Development subsystem. It allows you to define complex models of business processes using the three-dimensional extension of the BPMN model (Waszkowski, 2018). The use of 3D aspects allows you to place more information in the diagrams, such as data flows, analytical and statistical data, and information from the execution log of process instances.

Mathematical Model Integrator supports the integration of mathematical models into the process flow. Thanks to it, it is possible to solve optimization tasks in automatic tasks of a business process and to control them using process data. The calculation results can then be used to control the process flow on BPMN gates. The solution is available in the Aurea BPM platform.

10.3   METADATA REPOSITORY SUBSYSTEM

Figure 10.3 presents all modules of the Metadata Repository subsystem. It includes such modules as Processes, Tasks, Lists (or Registers), Scripts, Notifications, Dictionaries, Document Templates, and Filters.

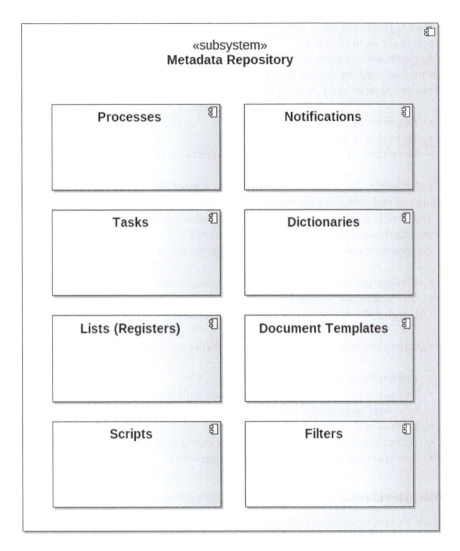

FIGURE 10.3 Metadata repository subsystem.

The process model management module (processes) is responsible for the following:

- Storing process definitions modeled in the BPMN Process Modeler
- Managing access to process definitions in accordance with user privileges
- Managing process versions
- Storing documentation of processes

The tasks module is responsible for storing and editing task definitions in processes, data definitions, roles, and information about data visibility in the user interface.

Data registers are freely definable named sets of records which are stored in the system in the form of tables.

Registers are definable lists in the form of tables or more complex data structures that may be used for storing data used in business processes. The Register module allows you to store and edit the following:

- Definition of registers
- Definition of data structures
- Definitions of events related to the manipulation of given registers

It is also responsible for managing access to register definitions and data structure definitions and managing the versions of registers and data structures.

The Scripts management module (Scripts) allows adding, updating, and deleting script definitions, as well as executing the scripts.

The Notifications module is responsible for the following:

- Configuring notifications depending on the occurrence of specific events
- Configuring the actions to be performed when certain events occur
- Logging the occurrence of events in the system

The Dictionaries module is responsible for storing the following:

- Definitions of internal dictionaries
- Definition of external dictionary sources with access via REST API, database link, custom API, etc.
- Definition of dictionary data validation
- Definitions of events related to the manipulation of dictionary data
- Configuration of access rights to dictionary data

10.4   DEPLOYMENT SUBSYSTEM

Figure 10.4 presents all modules of the Deployment subsystem. It combines the following modules to support end-user application generation: Application Generators, Data Model Generator, System Configuration Generator.

The Application Generators module groups together programs for generating end-user applications to handle tasks. These include such programs as user interface generator, business logic generator, forms generator, registry generator, report generator, etc.

Data Model Generator is the module responsible for transforming data models into database objects.

The System Configuration Generator is responsible for generating the system configuration for deployment in a specific customer environment. It allows configuring system roles, system functionalities, menus, dashboards, and access rights.

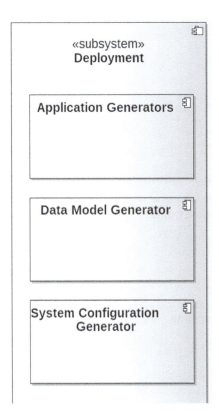

FIGURE 10.4   Deployment subsystem.

10.5   RUNTIME SUBSYSTEM

Figure 10.5 presents all modules of the Runtime subsystem. It includes such modules as Process Engine, Business Rule Engine, Forms Runtime, Lists and Registers Runtime, Runtime Database, Scheduler Service, Optimization Problem Solvers, and Simulation Runtime Interface.

The business process engine (Process Engine) is responsible for running process instances on the basis of BPMN models stored in the repository of process models (Metadata Repository, Processes).

Business Rule Engine, on the basis of business process data, determines the output values of business rules defined in Business Rules module (Design and Development) and stored in Tasks module (Meta-data Repository).

The Forms Runtime environment is used to prepare screen forms for display in the user interface (Brandl, 2002). Form definitions are taken from the UI Designer and Process Data Modeler modules (Design and Development subsystem), the data

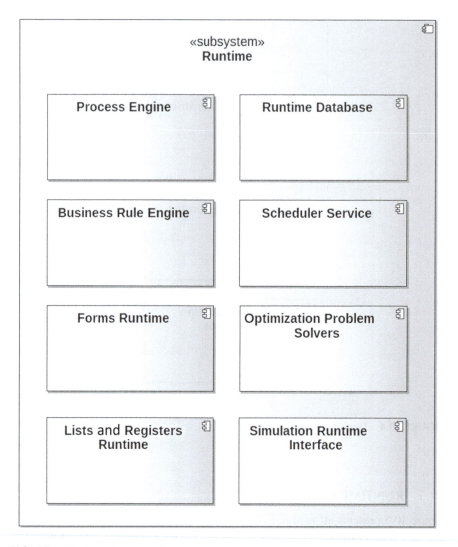

FIGURE 10.5   Runtime subsystem.

of a particular process instance are read and written in the Tasks module (Metadata Repository subsystem). In addition to data, forms also contain Task Actions, standard, and custom. Standard actions include the following:

- Task submission
- Task delegation
- Task escalation
- Task suspension
- Task consultation
- Task cancellation

- Process suspension
- Process cancellation

Custom actions are defined together with the task definition in the UI Designer (Design and Development) module. These are actions specific to a given task in a specific business process, for example, Print a report, Recalculate payments.

The Registers Runtime allows you to manage registers and the content of registers. Registers are defined in the Design and Development subsystem. The register definition data is stored in the Registers module (Metadata Repository), while the data of the registers themselves are stored in the Runtime Database module.

The Runtime Database stores all data resulting from the operation of processes, tasks, registers, and business rules in the runtime environment of the system. This data is available to the end user. The data stored in the database include the following:

- Data of running business process instances
- Instance data of running tasks
- Content of data registers
- Content of dictionaries
- Documents created by the end user

The aim of the Scheduler Service is to determine the optimal task execution order in the business process and to allocate appropriate resources to perform these tasks. The Optimization Problem Solvers module is an interface to the Problem-Solving Environment (PSE).

It allows you to formulate optimization tasks and embed them in business process models. Thanks to this, during the execution of a business process, in addition to determining the value of parameters on the basis of simple formulas and business rules, it is possible to solve complex mathematical problems and use the results to set parameters and, consequently, control the course of the process. The module's task is to store the definitions of mathematical models as well as algorithms and methods of their solution, and then at the right moment to call external solvers with appropriately set input data for calculations based on the data of a specific process instance.

The Simulation Runtime Interface allows you to simulate the execution of one or more business processes. This module has a user interface that provides access to all saved process data changes. Thanks to this, after performing a series of real or simulated processes, it is possible to recreate each of them step-by-step and observe changes in the values of selected parameters (After-Action Review).

10.6   LIFETIME MANAGEMENT SUBSYSTEM

Figure 10.6 presents all modules of the lifetime management subsystem. It includes such modules as Version Control, Change Management, Release Management, and Documentation Management.

Lifetime management subsystem is responsible for the administration of a system from provisioning, through operations, to retirement. Every IT system, resource, and

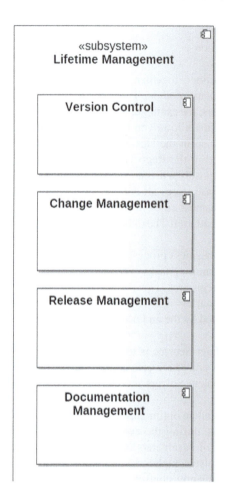

FIGURE 10.6 Lifetime management subsystem.

workload has a life cycle that has to be controlled and managed. In LCDP plat-forms, the life cycle management subsystem consists of the following modules: Version Control, Change Management, Release Management, and Documentation Management (Mrugalska, Tareq, 2017).

Version Control module is responsible for managing changes to models, scripting code, documents, templates, and other information necessary to produce a system in the end-user environment. It helps developers and citizen developers to store and manage results of their work related to design a system using LCDP.

Change management module is responsible for conducting the process of introduc-ing changes to the system in a controlled and coordinated manner. It reduces the pos-sibility that unnecessary changes will be introduced to a system without forethought, introducing faults into the system or undoing changes made by other users of software.

Release management module is responsible to control the process of converting models and design to a final product, which is the IT system deployed in the runtime environment (Jasiulewicz-Kaczmarek, 2018).

Documentation Management module is responsible for storing system documentation. You can also use this module to record document-related events such as development, approval, update, version control, and delivery to end user and project team.

Documentation is an important part of software engineering and includes types of documentation such as system requirements, software architecture and design, technical documentation, such as code, algorithms, interfaces, and APIs, as well as manuals for the end user, system administrators, and maintenance.

10.7   MANAGEMENT INFORMATION CONSOLE SUBSYSTEM

Figure 10.7 presents all modules of the management information console subsystem. It includes such modules as Dashboards, Reports, Key Performance Indicators, Business Analytics, 3D Analytics, Version Control, Change Management, Release Management, and Documentation Management.

FIGURE 10.7   Management information consoles subsystem.

A management information system (MIS) is used for decision-making, and for the coordination, control, analysis, and visualization of information.

In a corporate, the goal of the use of a management information system is to increase the value and profits of the business. This is done by providing managers with timely and appropriate information allowing them to make effective decisions within a shorter period.

In LCDP platforms, the life cycle management subsystem consists of the following modules: Version Control, Change Management, Release Management, and Documentation Management.

10.8   SECURITY SUBSYSTEM

Figure 10.8 presents all modules of the security subsystem. It includes such modules as Identity Management, Access Control, Auditing, History Management, and Encryption.

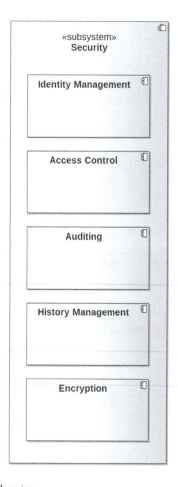

FIGURE 10.8   Security subsystem.

Security subsystem allows for the identity management and access control for both the users that model processes and prepare data entry forms and the end users. Furthermore, the subsystem supports data change auditing at the database level and history management at both application and database levels. It also allows for data encryption when the system concerns sensitive data.

10.9   SYSTEM ADMINISTRATION SUBSYSTEM

Figure 10.9 presents all modules of the system administration subsystem. It includes such modules as Users and Groups, Localization, Configuration, and Import/Export.

System administration subsystem is responsible for administration of users and user groups. It also supports user interface localization for different languages. Furthermore, it maintains all the configuration variables for the whole system. It also allows for data import and export.

10.10   PERFORMANCE AND SCALABILITY SUBSYSTEM

Figure 10.10 presents all modules of the performance and scalability subsystem. It includes such modules as Performance Analyzer, Queues, Load Balancer, and Asynchronous Task.

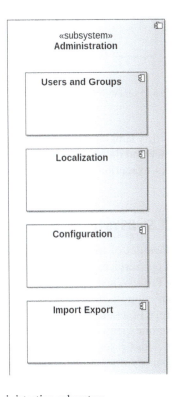

FIGURE 10.9   System administration subsystem.

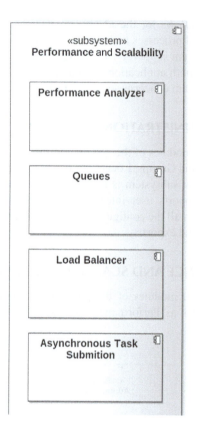

FIGURE 10.10   Performance and scalability subsystem.

Performance and scalability subsystem supports database and application tuning in terms of user interface and data processing performance. It allows for load balancing of process tasks execution, queues management, and asynchronous task execution.

10.11   MAINTENANCE SUBSYSTEM

Figure 10.11 presents all modules of the maintenance subsystem. It includes such modules as Application Log Service, Database Log Service, System Resource Monitoring, Issue-tracking Process, and Troubleshooting.

Maintenance subsystem supports the process of detecting and removing of existing and potential errors in a software code or process definitions that can cause the system to behave unexpectedly or crash. It includes both database and application logs. It also includes resource monitoring as well as troubleshooting and issue tracking functionality that can be used by both end users and system designers.

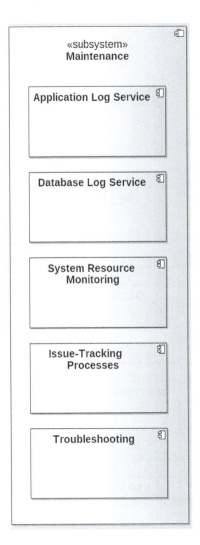

FIGURE 10.11   Maintenance subsystem.

10.12   INTEGRATION SUBSYSTEM

Figure 10.12 presents all modules of the Integration subsystem. It includes such modules as SOAP Web Services, REST API, Database API, MS Office Integration, ERP Integration (e.g., SAP).

The purpose of the integration subsystem is to deliver a vast set of different integration modules for various applications that might be used in the implementation environment. The integration involves the use of different techniques that allows interoperability such as Web services, APIs, database integration, plugins, and others.

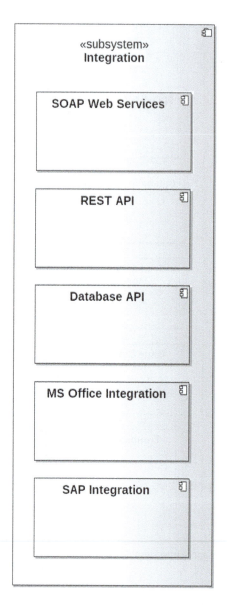

FIGURE 10.12   Integration subsystem.

10.13   CONCLUSIONS

This chapter presents all modules that make up the Aurea BPM Low-code Development Platform. They constitute the substantive layer of the platform and are responsible for creation, installation, and launching of applications in the end-user environment. In addition to the described subsystems, Design and Development, Metadata Repository, Deployment, and Runtime, the subsystems of the system management layer are also

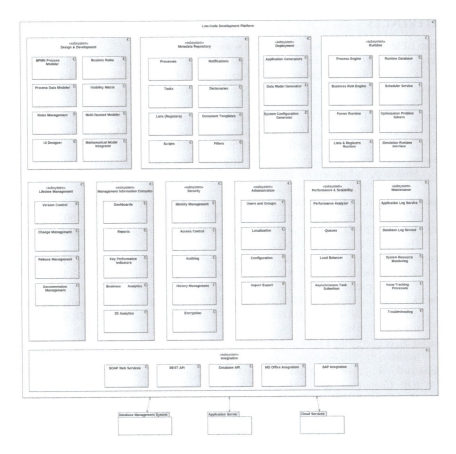

FIGURE 10.13   Detailed system architecture.

used in the Aurea-BPM Low-code Development Platform as support for the process of system development and maintenance (Aurea, 2022). The layer consists of the following subsystems: Lifetime Management, Management Information Consoles, Security, Administration, Performance and Scalability, and Maintenance.

The use of the architecture described in this chapter allows separating the system development subsystems from the runtime subsystems while maintaining a close relationship between the models of the designed application and its runtime form. This approach significantly speeds up the introduction of changes to the software and gives the opportunity to manage changes to the implemented and used system on an ongoing basis.

The detailed system architecture is shown in Figure 10.13. It shows all the modules and subsystems that make up the low-code development platform.

ACKNOWLEDGMENT

This work was supported in part by the grant UGB 22–799 financed by Military University of Technology, Warsaw, Poland.

REFERENCES

Aurea BPM System Documentation (2022) www.aurea-bpm.com.

Brandl A. (2002) Concepts for Generating Multi-User Interfaces Including Graphical Editors, *Computer-Aided Design of User Interfaces III*, DOI: 10.1007/978-94-010-0421-3_15.

Braude E.J., Bernstein M.E. (2011) *Software Engineering: Modern Approaches*, John Wiley & Sons.

Guerrero C.V.S., Lula B. (2002) A Model-Guided and Task-Based Approach to User Interface Design Centered in a Unified Interaction and Architectural Model, *Computer-Aided Design of User Interfaces III*, DOI: 10.1007/978-94-010-0421-3_11.

Jasiulewicz-Kaczmarek M. et al. (2018) Implementing BPMN in Maintenance Process Modeling, *Advances in Intelligent Systems and Computing*, Volume 656, DOI: 10.1007/978-3-319-67229-8_27.

Kiedrowicz M. et al. (2016) Optimization of the Document Placement in the RFID Cabinet, *MATEC Web of Conferences*, Volume 76, DOI: 10.1051/matecconf/20167602001.

McKendrick J. (2017) *The Rise of the Empowered Citizen Developer*, Unisphere Research, a Division of Information Today Inc.

Mrugalska B., Tareq A. (2017) Managing Variations in Process Control: An Overview of Sources and Degradation Methods, *Advances in Ergonomics Modeling, Usability & Special Populations*, pp. 377–387, DOI: 10.1007/978-3-319-41685-4_34.

Nowicki T. et al. (2016) Data Flow Between RFID Devices in a Modern Restricted Access Administrative Office, *MATEC Web of Conferences*, Volume 76, DOI: 10.1051/matecconf/20167604004.

Sanchis R. et al. (2019) Low-Code as Enabler of Digital Transformation in Manufacturing Industry, *Applied Sciences*, Volume 10, DOI: 10.3390/app10010012.

Waszkowski R. (2018) Multidimensional Modeling and Analysis of Business Processes, *IOP Conference Series-Materials Science and Engineering*, Volume 400, DOI: 10.1088/1757-899X/400/6/062031.

Waszkowski R., Bocewicz G. (2022) Visibility Matrix: Efficient User Interface Modelling for Low-Code Development Platforms, *Sustainability*, Volume 14, No. 13, p. 8103. DOI: 10.3390/su14138103.

Waszkowski R., Kiedrowicz M. (2015) Business Rules Automation Standards in Business Process Management Systems, in: *Information Management in Practice*, Faculty of Management, University of Gdańsk.

Woo M. (2020) The Rise of No/Low Code Software Development—No Experience Needed?, *Engineering*, Volume 6, DOI: 10.1016/j.eng.2020.07.007.

11 Monitoring and Improvement of Data Quality in Product Catalogs Using Defined Normalizers and Validation Patterns

Maciej Niemir and Beata Mrugalska

CONTENTS

11.1 CONCEPT OF PRODUCT DATA QUALITY

Data quality is most easily defined as "fitness for use" (Wang and Strong 1996; Batini et al. 2009). If a set of data used under specific conditions meets all the implied and stated requirements of users, it can then be said to be (in the sense of ISO 25012) of high quality. The concept of data quality can be decomposed in many ways, examining its characteristics from different perspectives. In this concept, attributes—metrics of data quality—are used to operationalize the dimensions of data quality and identify data for measurement as measurement point, measurement technique, and measurement scale (Batini et al. 2015). Measuring data quality is the central issue of many scientific studies. Over the past few decades, a number of methodologies have been developed, with frameworks defining from several to dozens of dimensions. A comprehensive review of these was presented by Cichy and Rass (2019). Many studies also address procedures and techniques for measuring data quality: using interviews and surveys (Price, Neiger, and Shanks 2008) or validation rules (Fan et al. 2008). Models of procedures and analysis techniques have also emerged for the process of identifying data defects and determining data quality metrics in a specific context (Gelinas, Dull, and Wheeler 2014; Heinrich and Klier 2009). Data quality in terms of product data, commonly understood as data that includes all product-related information that can be read, measured, and structured in an appropriate format (Kropsu-Vehkapera and Haapasalo 2011), fits completely into these definitions and area of research. Product data must be accurate, complete, consistent, timely, valid, and reliable, which is a key element in companies' business processes. Errors in this area can significantly affect the financial performance of companies, especially in terms of online sales, as confirmed by numerous studies (Liu et al. 2019; Wilda Kurnia Putri and Vera Pujani 2019).

11.2 CONCEPT FOR IMPROVING DATA QUALITY IN THE ELECTRONIC CATALOG DATABASE

This study focused on finding ways to improve the quality of product data by selecting and testing the effectiveness of appropriate standardization algorithms and validation rules that verify the process of adding and updating product data. The idea was to find ways to instantly and automatically inspect product data entered into the database, without the involvement of a controller person verifying the information. This is critical to reduce data management costs, especially with large volumes of data filled in by people without proper training or data collected from uncertain sources. This is the case when it comes to databases filled by communities or groups of companies working toward a common tool such as price comparison sites, marketplace platforms, and information brokers. The research was conducted on the basis of observation of an existing product database storing information on 41 million products of manufacturers operating in the Polish market across the industry spectrum. Manufacturers populate the data themselves, in an information system that has been in operation for many years and has numerous security and validation methods implemented. The research was also supported by the experience gathered during the implementation of a prototype product data aggregator combining more than 50 different data sources with different formats and structures. It should be noted that validators that limit data entry are not an innovative way to correct data, given that they were created with the advent of the first IT systems for data management. The value of this research is to select and test the usefulness in real-world conditions of specific, unusual validation rules, because in the author's experience, actual implementations of validators in IT systems are usually limited to protecting those systems from failure, unhandled exceptions, etc. Often their control is narrowed to testing the length of text strings and the limits of numeric values. Data consistency and accuracy remain the responsibility of product data management specialists, but such individuals are not always aware of the capabilities of the information systems that can help them. Furthermore, less experienced people who have not worked with large volumes of crowd-managed data may not anticipate how the human factor can affect the quality of collected data.

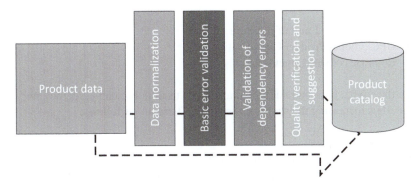

FIGURE 11.1 Division of standardization and validation methods.

Source: Own source.

In implementing validation rules, it is important to plan the order in which they will be triggered during the process of adding or editing data, so it is worth doing this at the planning stage. It is also worth to acknowledge the existence of certain types of mechanisms that can be used, and then ordering them in the right sequence. This concept divides validation methods into four main areas (Figure 11.1):

(1) **Data normalization**, where data is subject to ordering and editorial correction, with no change in the meaning of the given content of a given attribute/field.
(2) **Validation of fundamental errors**, where data are subject to verification for correctness with the accepted assumptions, standards, and norms for the content of a given attribute/field. These are both technical and logic-based errors.
(3) **Validation of errors resulting from specific dependencies**, where information from other fields/data sources is needed to assess the correctness of the data. These are contextual, logical errors.
(4) **Quality verification and suggesting changes**, understood as verification that does not give certainty that an error has occurred, or gives such certainty, but the data is conditionally allowed. The system in such a case can indicate the probability of an error and give suggestions to the inputter, or give the appropriate information to the controller, and then such validation will not block the input. In this space, there is room for data validation that takes into account statistics and artificial intelligence.

The range of data used for the study takes into account the so-called basic data about the product described in the publication (Niemir and Mrugalska 2021):

- GTIN (Global Trade Item Number)
- Label description/Product name
- Brand name

- Product image
- Net content and unit of measure
- GPC (Global Product Classification)
- Countries of sale/target market

Since the source base had additional attributes used in e-commerce, they were also added, thus increasing the scope of the study. These are as follows:

- Product webpage
- Marketing description of the product
- Product description language (a dictionary field specifying the language for both the product name and marketing description)
- SKU (Stock Keeping Unit)—the product's internal symbol, assigned by the manufacturer

11.3 DATA NORMALIZATION

Analysis of the data began with a search for elements that would correct the data editing process, according to the concept. Such activities include improving the notation of the link to the product page—that is, normalizing the URI (Uniform Resource Identifier) according to RFC 3986, where each link must start with a protocol, etc. But normalization also includes editing errors in text data, which can be corrected. Errors in this area can result from both human factor, where a system user may by mistake enter, for example, a double space between words, or from the peculiarities of external information systems with which data is synchronized. For example, systems, when exporting data, protect the quotation mark by adding a backslash character ("\") in front of it, or multiplying it, to distinguish such a character from marking the beginning and end of a string, while systems that import data may not convert such a value without setting additional parameters by an unaware user. There are also situations where systems do not display hidden characters, but write them to the database (e.g., hidden characters marking the beginning of a new line, tab characters), or those where the characters simply cannot be seen on the screen (e.g., spaces at the end of a string). In the author's experience, computer systems usually do not normalize text-related fields when saving data, and if such normalizations have already been implemented, they are severely limited, for example, to built-in commands that remove unnecessary spaces at the beginning and end of a text string. The problem is that such commands take into account regular spaces, and not, for example, hard spaces, which appear the same on the screen, but are written as a different digital code, which means that traditional space removal algorithms will not work. In order to increase the effectiveness of the developed algorithms—validation rules, the study used the commonly used Unicode character encoding standard and normalized data by referring to defined groups of characters, including lowercase letters, uppercase letters, special characters, etc. Such groups take into account all diacritical marks, and, for example, various forms of spaces. When building validation rules, attention was also paid to the fact that, depending on the field, the algorithm cannot remove some characters, but must replace them with others. For example, the beginning-of-line character (enter) in single-line fields is better replaced by a space, because otherwise there may be a concatenation of two adjacent words. On the other hand, a tab character replaced by a space can result in two spaces next to each other. For this reason, it is important to maintain the correct order of use of normalizers after each other, so that the next one removes a possible error created by the operation of another normalizer.

The entire set of available data was used for the study, that is, *41.49 million* product data. Using created normalizers, *9.5%* of them, *3.39 million*, were corrected. A total of *3.94 million* corrections were made. A detailed list of the normalizers identified, along with their impact on improving the database, can be found in Table 11.1.

As part of the normalizers for digital fields, such as net content, it is also worthwhile to perform a conversion of the notation of multi-digit numbers. Depending on the country, the decimal separator can be a period, comma or momayyez. Three-digit group separators are also used to separate thousands, millions, etc., and sometimes this is an unbroken space, sometimes a comma, and sometimes a period, so the lack of conversion can be the cause of a skewed recording of values. However, in

TABLE 11.1
Normalizers Used in the Tested Product Base

Description of Operation	Call Sequence	Impact
Removal of multiple spaces and hard spaces ("Space separator" [Zs]) inside the text in the text attributes examined	6	2.8M
Removal of multiple consecutive newline start characters ("enters") and normalization to the Linux standard (LF) in the marketing description field. Pressing the "Enter" button is encoded differently on Linux, Windows, and Mac systems. Standardization of the URI of the link to the product website	5	381k
Standardization of the URI of the link to the product website	insignificant	295k
Removing control characters from text fields, whereby the control character can be any of the 65 characters of the "Control" [Cc] group of the Unicode code list except tabs and enters (\t\n\r)	3	185k
Decoding html entities (in order to properly display some characters in html, you need to encode them, e.g., displaying the "<"character requires encoding to "<" or "<") into valid letters	1	89k
Replace the start-of-new-line character ("enter") with a space for single-line text fields (product, brand and SKU names)	4	70k
Removal of empty spaces at the end of text (from 17 Unicode characters of the "Space separator" [Zs] group)	9	62k
Replacing the tab character with a space in all text attributes	2	34k
Removal of blank spaces at the start of the text (from 17 Unicode characters of the "Space separator" [Zs] group)	8	8.2k
Reduction of duplicate quotation marks	7	7.9k

the product database tested, this problem did not occur, and therefore is not included in the table.

The normalized values of product data attributes in the tested database were subjected to further testing according to the aforementioned concept, i.e., the process of validating basic errors. Starting with normalization was crucial, because if the order of activities performed was different and started with data validation—testing the obligatory fields—the product name could contain only a few spaces and such data would meet the criteria.

11.4 BASIC ERROR VALIDATION

At the basic error validation stage, the following is checked:

- Whether mandatory fields have been filled in
- Whether the minimum and maximum values in the fields have not been exceeded
- Whether all specific fields comply with implemented rules, standards, and technical specifications.

Unfortunately, in the case of product data, respected norms and standards do not exist for all attributes, which makes it difficult to build effective constraining validators. For this reason, when designing validators, assumptions should be made that are compatible with most of the solutions respected by external companies with which subsequent information exchange will take place. This is contrary to common practice, where the size of the text field only limits the possibility of storing data in the database, not the limitations of the market. The same is true for taking care of mandatory fields. It should be remembered that a liberal approach to data completion generally ends up with missing data and, consequently, later problems with data synchronization. Therefore, it is worth taking care of the tightness of the base by entering many mandatory fields. A detailed comparison of guidelines for the correct entry of product data in selected e-commerce platforms is described (Niemir and Mrugalska 2021, 2022).

One example of a technical validator, quite often implemented in IT systems, is the verification of the check digit of a GTIN (Global Trade Item Number)—a unique identifier of a product in accordance with international standards. The GTIN is constructed in such a way that its last digit regardless of the type and length of the number is calculated by the same formula:

$$S = \sum_{i=1}^{n-1}((2-(-1)^i)\cdot d_i) = 3d_1 + d_2 + 3d_3 + d_4 + \ldots + ((2-(-1)^{n-1})\cdot d_{n-1})$$

where d_i is the next digit of the code, and n is the length of the code.

The check digit will be the value by which the resulting sum must be topped up to make it divisible by 10.

$$c = -S \bmod 10$$

FIGURE 11.2   Anatomy of GTIN-13.

Because of the implementation of such a validator, fewer errors are encoun-
tered due to incorrect entry of the GTIN into the database, but they still happen.
Unfortunately, such a validator is not sufficient to assess the reliability of the number,
so it is worth using a set of validators for this field to be sure of the correctness of it.
Validation should start by verifying the length and type of the field. The GTIN number is always a number. The general numeric range is 10,000,007–99,999,999,997 (or
9,999,999,999,994 if the system assumes only storing numbers available at retail).
The input format of the data should be a string of digits of length 14, 13, 12, or 8. The
length of the string defines the type of GTIN: GTIN-14, GTIN-13, GTIN-12 (U.S.
UPC numbers), GTIN-8. Any number of a lower type can be presented in a higher
type by adding an insignificant zero to the left of the string. One of the most important validators for verifying the correctness of a GTIN, and one that is often not
implemented, is the verification of the country prefix—a three-digit number located
at the beginning of each GTIN-13, GTIN-12 and GTIN-8. In GTIN-14, this number
is located right after the digit that identifies the packaging level. Figure 11.2 presents
the structure of the most common retail GTIN-13 number, with an indication of the
occurrence of the prefix.

The country prefix is a dictionary, available in the form of a table on the web
pages of GS1 organization, which manages the issuance of codes and ensures their
uniqueness throughout the world. The dictionary is a subject to a change in case a
new prefix is added but this happens relatively rarely. The table contains information
on both the prefixes used in a given country, as well as the reservation of prefixes for
specific applications, to which retail products cannot be assigned, or can only be used
internally, as the numbers can duplicate globally.

Another validator in the context of a GTIN, more difficult to implement, is the
company prefix validator. The GS1 organization does not publish lists of such prefixes, but provides cloud-based services for their verification (e.g., the GEPIR system,
Verified by GS1 project), so IT systems to verify the validity of such numbers must
connect to them on the fly or in batches.

To summarize the issues of GTIN validation—there are many validators that com-
plement each other. Numerous tests have been conducted in this regard on dozens of

databases from various sources. It turns out that in many cases there is more often a problem with the incorrect country prefix and the length of the number itself than with the verification of the check digit, and the error rate is up to several percent across the entire product base of a given database. In 2017, a study was conducted to verify the correctness of company prefixes on verified GTINs from the point of view of country prefix and check digit, for 50 external data sources from information brokers, stationary, and online stores. The data was placed in a common product aggregator. The study found that for more than 3.5 million product data collected, 3% of the numbers were illegal (no such company prefixes existed), and 2% should not have been available for sale.

The database tested, had all the validators relating to GTIN numbers, as well as restrictions on the maximum lengths of product attribute fields, and fit into the field restrictions proposed by market representatives described in the aforementioned publications. Despite this, the possibility of adding other validation elements was analyzed in order to increase control over the data in the database. As a result of the analysis, more than 21 million defects were detected in attributes that classified product data for rejection. Most of the defects were attribute-specific, but some could be successfully run for several fields.

Table 11.2 presents the results of the common validators, along with the number of occurrences of errors in each field. A total of 713,000 errors were detected this way. There were three main groups of problems. The first problem concerned the way product data was written into text fields. It was detected that instead of the correct information, dots alone were being entered into the fields in order to bypass the no-fill protection. Sometimes the data began with hyphens, and sometimes all the text was in extra quotation marks or apostrophes. This error was eliminated by verifying the first character of the text string, which, according to the relevant rule, must be an element from the group of letters or digits of the Unicode table.

The second type of error concerned the conversion of the character set. Nowadays, the aforementioned Unicode format is widely used—a computer standard for encoding a character set that includes the letters of most of the letters used in the world, as well as symbols, emoji and formatting codes. Unfortunately, in the past there were many such formats, each of which encoded characters differently, so inadequate conversion from an older computer system to today raises many problems with the correct display of characters. Instead of the correct element, the screen then shows frame fragments, question marks, completely different letters, or sets of two different letters. Detection of such characters is crucial to maintain good quality data sets.

The third type of problem involved technical issues and was a consequence of importing data from spreadsheets, where the data is a calculated element of another field or data source.

From the perspective of detected number of errors, individual validators, created for a particular field, provided the most benefit. Table 11.3 presents a detailed list of proposed validators, along with a demonstration of effectiveness.

The most common errors concerned duplicate product names (15.3 million), which, in the understanding of the name as a unique description of the text on the label, is a significant error. Often there were generally short names in the database,

TABLE 11.2
Common Basic Error Validators

Proposed Validator	Product Name	Brand Name	SKU	Marketing Desc.
The text in the field does not start with a letter or number (is not a character from group N and L of the Unicode table)	320k	41k		126k
There are characters in the text that indicate errors in the conversion of code tables (Unicode, Windows 1250, ISO/IEC 8859–2)	151k	14		2.8k
The text in the field is a saved Microsoft Excel formula. Instead of data, for example, the VLOOKUP command appears	1.4k	5.4k	14k	5.1k

TABLE 11.3
Individual Basic Error Validators

Proposed Validator	Product Data Attribute	Impact
Duplicate product name	Product name	15.3M
Product name contains only one word	Product name	3.4M
Multiple underscores in the product name	Product name	151k
Brand name has too many words (more than four words containing more than two characters each)	Brand name	1.1M
Prohibited brand name (dictionary list)	Brand name	163k
The brand name is the same as the product name	Brand name	37k
Incorrect website status	Product webpage	396k
Prohibited product website domain (dictionary list)	Product webpage	54k
Website under construction	Product webpage	3.7k
Photo is too low resolution (photo smaller than 320×320)	Product image	5.5k
Photo is too high resolution (resolution greater than 64 Mpix)	Product image	158

e.g., "blouse," "bread," which indicated that people entering data into the database confused this field with the "common product name" field. Therefore, a validator was also proposed that examines the product name to see if only one word is entered in the field. The validator perfectly captures common names, as well as names that only contain the product code. The underline validator is a technical validator that eliminates noted errors regarding common underline characters in texts. Presumably, these are data incorrectly converted by some external system.

The brand name validators examined whether the name was the same as the product name and did not appear in a list of blocked names. The list was defined manually and included names such as "no name". Experimentally, it was also suggested to limit the number of words in the brand name due to the fact that this field was often incorrectly filled in and contained the full company name or simply too much content.

For the product's Web pages, a set of validators has been proposed that examines whether the page exists (returns the appropriate HTTP code), whether it is under construction (this is detected on the basis of defined patterns), and whether it refers to forbidden domains (e.g., does not redirect to the main page of a news site, a service provider's site, etc.).

For product image validators, the minimum and maximum dimensions needed to be met assuming minimum compatibility with Google Merchant Center were assumed (Niemir and Mrugalska 2022a).

11.5   VALIDATORS OF DEPENDENCY ERRORS

When analyzing this group of errors, we must keep in mind all attribute values that depend on the selection of information from other fields. In the analyzed database there is one such direct dependence—it is the dependence of the unit of measure on the net content. There are several units of measure in the list, such as a piece, which should have content restrictions only to a natural number. Such a case was taken into account when creating another validator. The validator detected over 687,517 products where the net content was a fractional number, sometimes even equal to 0.00001 piece.

Similar relationships can be built based on the information content of one field relative to another. For example, in the product name (note that this is not the common name of the product, but a name describing the product, otherwise known as label description), you can require information about the brand name, or the net content, and then verify the content of these fields against each other. This will not eliminate all errors related to the correctness of writing the product name, but it will partly help to maintain important rules for its construction and take care of data consistency. In e-commerce, including the brand name and net content in the product name is an unwritten standard (Niemir and Mrugalska 2022b). Unfortunately, in the tested database, the vast majority of product names did not have such information, so it was decided to omit verification of this type of validator.

11.6   VERIFYING QUALITY AND SUGGESTING CHANGES

This type of validation is unusual in the way it works, as it allows data to be written to the database even though an error is present, or the data is likely to be incorrect. Depending on the implementation of this type of validation, the occurrence of an event before the data is written may inform the data inputter of a possible error made, suggest changes to improve quality or even suggest an exact change, inform the data inputter of standards that have not been met at this point, or suggest the consequences of not taking action. Such an event can also be the start of a parallel process outside of data recording, such as informing the controller of potential data errors, or influencing a reduction in the aggregate data quality score, which as an indicator can be presented later in reports. It can also be part of a gamification system designed to introduce competition among inputters.

The simplest validator that can be assigned to this type of validation is the obligatory validator for optional fields. For example, if a recommended but optional product photo is not entered, its absence is recorded and reduces the quality of the entire product description, while the data will be saved.

Some of the validators assigned to this type of validation may be based on rules for which it has been found that they do not apply to 100% of cases, but mostly work. In that case, you cannot downgrade the quality of the data, but you can inform the person entering the data about the probability of making an error. Similarly, such a rule can be applied to validators that result in a probability of effectiveness at a certain level. Effectiveness thresholds can be used to accept, reject, or inform about a given fact.

Table 11.4 shows the detailed results of the analyses of the selected validators along with the impact on the product base under study.

For the attribute specifying the product name, two validators were selected and tested. It turned out that almost 15 million names are written using only capital

TABLE 11.4
Validators to Verify Quality and Suggest Changes

Proposed Validator	Refers to Attribute	Impact
Product name is written in capital letters	Product name	15M
Product name starts with a lowercase letter	Product name	1.4M
Brand name starts with a lowercase letter	Brand name	2.1M
Brand name used only once in the entire database	Brand name	100k
Brand name consists of less than two letters or numbers	Brand name	5.7k
Too few words in the marketing description (less than six)	Marketing description	11.4M
Too few characters in the marketing description (less than 20)	Marketing description	6.3M
Marketing description starts with a lowercase letter	Marketing description	4.2M
Marketing description described in capital letters	Marketing description	2.7M
Incorrectly indicated description language	Language description	174k
Temporary GPC brick code 99999999 is indicated	GPC	762k
Web page does not contain GTIN information	Product webpage	5.8M
Net content of product out of scale	Net content	3.9M
Net content of the product probably out of scale	Net content	1.8M
Incorrect net content unit for the selected GPC category	Net content	50k
Insufficient product photo dimensions for all applications (photo is $\geq$ 0.3 Mpix, but dimensions do not exceed 900 × 900 pixels)	Product image	298k
Incorrect background of the photo (the background is not white)	Product image	247k
Photo of too low resolution (the photo is < 0.3 Mpix in size, but its dimensions exceed 320 × 320 pixels)	Product image	6.4k

letters, and more than a million begin with a lowercase letter. The content of the attribute itself may be completely correct and thus acceptable for writing, while the quality deviates from the accepted standards. This is a good example of the described idea of validation.

For the brand name attribute, it was examined, as mentioned earlier, whether the content begins with a lowercase letter. Furthermore, the number of relevant characters in the field was verified, and after normalization, 5702 brands were found to lack a minimum of two letters or numbers in the text. During the implementation of the tests, it turned out that some brand names that were filtered by this validator were nevertheless correct, while the vast majority were not. For this reason, it was decided to place this validator in the current group. Another validator tested was one that measures the frequency of use of a given brand name, since it was observed that users of the database enter other values into this field each time instead of the brand name. Here similarly—in most large collections, the brand should not occur only once, while there is no certainty that this is a bug. The validator must also be insensitive to the process of introducing new brands into the database.

Similar textual rules were applied to an optional field in the database called "marketing description". It was assumed that a correct and good quality description is one that has a minimum of 6 words, 20 characters, does not start with a lowercase letter and is not completely described with caps-lock. These are very basic, experimental assumptions that, as observed, have questioned the reliability of the data 24 million times anyway. One of the best ways to verify these types of fields should be artificial intelligence mechanisms that examine the content for relevant patterns depending on the product category, but training them well can be time-consuming.

For the first time in the list of validators, verification of the dictionary field specifying the language of description (for product name and marketing description) was proposed. During the base verification, it turned out that users are most likely to confuse the description language fields with the country of sale field, or the country of origin of the product. As part of the experiment, it was examined whether the indicated description language is other than Polish if there are Polish characters in the name or description. The result returned 173,592 errors, so it is reasonable to assume that extending the validator using appropriate natural language processing techniques—in this case, probably statistical verification of the occurrence of language-specific n-grams—should yield even greater results and signal a production implementation of such an extended solution.

Product classification is another attribute needed for detailed inspection. A quality verification was proposed to examine whether a so-called temporary classification was introduced instead of a specific value from a list. It was noted that this is often the choice of inputters especially when the classification is not popular. This is a typical validator that should influence the quality assessment of the base and trigger a series of additional events to help in the improvement process.

Storing information about a product's website link is a good way that can combine information stored inside any database, with distributed information stored directly by manufacturers, compare them and automatically update them. Experimentally, a mechanism has been set up to verify whether a pointed page stores information about a particular product, by looking for the GTIN number in the structural data of the page. The absence of such information (unfortunately, for nearly 6 million sites) is a signal of a lower quality index for such data.

The large volume of data tested also revealed a fair number of errors in entering net content with the unit, so it was decided to develop and test validators that would verify that the use of a given unit of measure with a given classification is appropriate, and that the reported net value for a given unit of measure with a given classification falls within a set range. Since it would have been too time-consuming to manually determine such relationships, the GPC-unit pair and the GPC-type of measure pair were measured against actual data in the database, taking into account the number of companies using such pairs and the number of products assigned to them. Thresholds were selected and tested for which units for a given GPC category could be considered correct or incorrect. Assignment of a pair: GPC—unit was considered incorrect when a minimum of 10 different sources (companies) had product data in the database in a given GPC category, while for a given type of measure there were less than 10% of the total and were entered by less than five different sources. The minimum and maximum net content thresholds were determined based on statistical data. The maxima and minima of net content for each unit, in each GPC item, for each company with active products were counted. The collected data was used to calculate the upper and lower medians. This created a conventional safe range of values. In addition, a margin of error was added, which was used to distinguish between a validator that determines a low and high probability of going off the scale. The impact of the validators built in this way proved to be high and is worth implementing. It was also noted that for each product GPC category, the graph showing the number of products for a given net content creates different lines with empty spaces inside. This can be clearly observed, for example, for the number of screws in a package, the weight of the contents of a spice bag, and the capacity of a beverage bottle. Future research therefore plans to use artificial intelligence algorithms to study anomalies inside the scale, not just at the extremes.

The last item examined was the product photo. Once again, photo sizes were examined, but this time to determine quality. Behind this proposed solution is the goal of informing the introducer that, in some cases, the photo may not be accepted by certain platforms (Niemir and Mrugalska 2022a). Background verification was also used to determine the quality of the photo. The algorithm takes measurements of the relevant areas of the photo and, based on these measurements, verifies that the background is equal and in what color.

11.7 CONCLUSIONS AND FURTHER RESEARCH

The validators proposed in the study provide some insight into the possibility of influencing data quality from the perspective of the IT system, in the process of filling and updating product data. The reported magnitudes of impact on the database under study make it possible to verify their effectiveness and prove the need for their implementation. Some of the validators proposed and analyzed in this study are experimental validators that require supervision (e.g., list of banned brands, list of banned web addresses, list of correct net content ranges for a given GPC) as well as deeper analysis (validators for GPC, images, language, net content, product web pages). These are just suggestions that give a clear picture of how the product base can be influenced to improve its quality. In the longer term, research into the feasibility of appropriate validators will be deepened with solutions mainly involving artificial intelligence (machine learning, computer vision) and natural language processing mechanisms (Muszyński, Niemir, and Skwarek 2022). This will make it possible to automate many elements, as well as solve problems not analyzed so far, such as follows:

1. Detecting typos in product names.
2. Detection of similar brands and prevention of duplication.
3. Detecting whether a product photo only shows a company logo.
4. Detecting whether a photo depicts a single product.
5. Detection whether a photo shows a product of a selected category.
6. Detection whether the photo contains promotional texts that interfere with the main product image.
7. Detection whether the photo depicts a product instead of product packaging.
8. Detection whether the photo is already used with other products.
9. Automatic verification that the data on the web page about the product is consistent with that posted in the database.
10. Detection whether the GPC category has been selected correctly. What is the correct category?
11. Detection whether the description language is consistent with the declared one. What is the correct one?
12. Automatic detection of net content anomalies.

REFERENCES

Batini, Carlo, Cinzia Cappiello, Chiara Francalanci, and Andrea Maurino. 2009. "Methodologies for Data Quality Assessment and Improvement." *ACM Computing Surveys* 41 (3): 1–52. https://doi.org/10.1145/1541880.1541883.

Batini, Carlo, Anisa Rula, Monica Scannapieco, and Gianluigi Viscusi. 2015. "From Data Quality to Big Data Quality." *Journal of Database Management (JDM)* 26 (1): 60–82. https://doi.org/10.4018/JDM.2015010103.

Cichy, Corinna, and Stefan Rass. 2019. "An Overview of Data Quality Frameworks." *IEEE Access* 7: 24634–24648. https://doi.org/10.1109/ACCESS.2019.2899751.

Fan, Wenfei, Floris Geerts, Xibei Jia, and Anastasios Kementsietsidis. 2008. "Conditional Functional Dependencies for Capturing Data Inconsistencies." *ACM Transactions on Database Systems (TODS)* 33 (2): 1–48. https://doi.org/10.1145/1366102.1366103.

Gelinas, Ulric J., Richard B. Dull, and Patrick Wheeler. 2014. *Accounting Information Systems*. Cengage Learning.

Heinrich, Bernd, and Mathias Klier. 2009. "A Novel Data Quality Metric for Timeliness Considering Supplemental Data." *ECIS 2009 Proceedings*. pp. 2651–2662. https://aisel.aisnet.org/ecis2009/14.

Kropsu-Vehkapera, Hanna, and Harri Haapasalo. 2011. "Defining Product Data Views for Different Stakeholders." *Journal of Computer Information Systems* 52 (2): 61–72. https://doi.org/10.1080/08874417.2011.11645541.

Liu, Aijun, Yan Zhang, Hui Lu, Sang-Bing Tsai, Chao-Feng Hsu, and Chien-Hung Lee. 2019. "An Innovative Model to Choose E-Commerce Suppliers." *IEEE Access* 7: 53956–53976. https://doi.org/10.1109/ACCESS.2019.2908393.

Muszyński, Krzysztof, Maciej Niemir, and Szymon Skwarek. 2022. "Searching for Ai Solutions to Improve the Quality of Master Data Affecting Consumer Safety." *Business Logistics in Modern Management Proceedings.* pp. 121–140.

Niemir, Maciej, and Beata Mrugalska. 2021. "Basic Product Data in E-Commerce: Specifications and Problems of Data Exchange." *European Research Studies Journal* XXIV (Special Issue 5): 317–329. https://doi.org/10.35808/ersj/2735.

Niemir, Maciej, and Beata Mrugalska. 2022a. "Product Data Quality in E-Commerce: Key Success Factors and Challenges." *13th International Conference on Applied Human Factors and Ergonomics (AHFE 2022).* https://doi.org/10.54941/ahfe1001626.

Niemir, Maciej, and Beata Mrugalska. 2022b. "Identifying the Cognitive Gap in the Causes of Product Name Ambiguity in E-Commerce." *Logforum* 18 (3): 9. https://doi.org/10.17270/J.LOG.2022.738.

Price, Rosanne, Dina Neiger, and Graeme Shanks. 2008. "Developing a Measurement Instrument for Subjective Aspects of Information Quality." *Communications of the Association for Information Systems* 22 (1): 3. https://doi.org/10.17705/1CAIS.02203.

Wang, Richard Y., and Diane M. Strong. 1996. "Beyond Accuracy: What Data Quality Means to Data Consumers." *Journal of Management Information Systems* 12 (4): 5–33. https://doi.org/10.1080/07421222.1996.11518099.

Wilda, Kurnia Putri, and Vera Pujani. 2019. "The Influence of System Quality, Information Quality, e-Service Quality and Perceived Value on Shopee Consumer Loyalty in Padang City." *The International Technology Management Review* 8 (1): 10–15. https://doi.org/10.2991/itmr.b.190417.002.

12 Simulation Exercises for the State Sanitary and Epidemiological Services Related to the Epidemic of Foodborne Diseases

Tadeusz Nowicki, Robert Waszkowski,
and Agata Chodowska-Wasilewska

CONTENTS

12.1   INTRODUCTION

Nowadays, simulation exercises are more and more common and are organized for various applications (Lateef, 2010; Cayirci, Marincic, 2009; Schirlitzki, 2007). Controlled training based on computer simulation has already achieved some standardization (IEEE Std 1516TM, 2010). In the case of an epidemic of foodborne diseases, sanitary inspectors are participants in training teams. Organized simulation exercises make it possible to practice the procedures and roles used by them during the epidemic. Simulation exercises should be organized to increase in complexity. Learning in a safe and managed environment provides essential hands-on experience that integrates key theoretical concepts with interactive, computer simulated situations. Controlled simulation-based training reduces risk as the learning

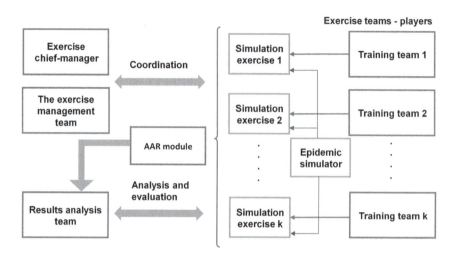

FIGURE 12.1   Structure of computer-based simulation exercises.

environments are safe, while sanitary inspectors can master the skills needed in a real-life situation. Simulation exercises organized today have a complex structure. There are not only exercise teams—players, but also other groups of participants necessary to carry out this type of exercise. It is worth remembering that each of the possible teams participating in the simulation exercise, including exercise preparation team, exercise management team, team for analyzing the results of exercises and system administrator (see Figure 12.1).

The chapter presents business processes related to the work of sanitary services in cases of an epidemic of foodborne diseases. On their basis, simulation exercises according to given scenarios were developed. The activities of the staff of the county-level State Sanitary Inspectorate were analyzed in terms of actions taken in crisis situations related to the emergence of a large food poisoning outbreak or an epidemic of foodborne infectious disease. The same analysis was carried out in the scope of actions performed by individual teams participating in the simulation exercises.

12.2 BUSINESS PROCESSES OF EPIDEMIOLOGICAL INVESTIGATION

Simulation exercises for sanitary services are similar to the anti-epidemic processes carried out during an epidemic of foodborne diseases. Application of business processes for different operations (Jasiulewicz-Kaczmarek et al., 2018), (Nowicki et al., 2016), (Waszkowski, 2018) and (Waszkowski, Kiedrowicz, 2015) in that example for the operation of sanitary inspectors during an epidemic of foodborne diseases specified at the highest level of detail is shown in Figure 12.2.

As a result of starting the process of epidemiological investigations, a sanitary inspector analyzes the cumulative number of reported cases. The result of the analysis is a decision of the inspector as to whether the reported cases concern an outbreak

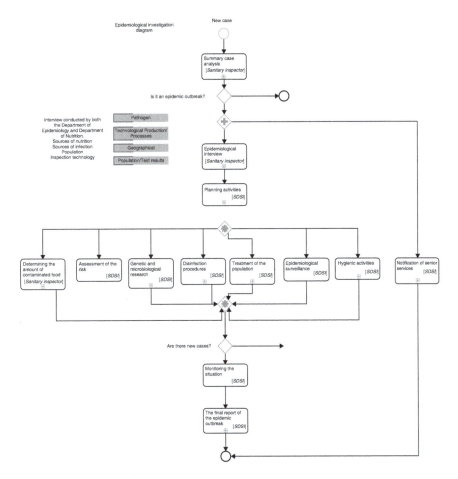

FIGURE 12.2 Epidemiological investigation business process diagram.

of disease or infection. In the event of an outbreak, further action is triggered in the investigation. Otherwise, the investigation ends. Further activities carried out within the investigation focus on the Sanitary Inspectors conducting interviews with patients and sanitary inspection in places where contaminated products may be found. Results and conclusions from interviews are recorded in the system based on the appropriate form of paper taken from the system. After completing interviews, sanitary officers plan further actions that should be executed as part of the epidemiological investigation.

The following activities are executed:

- Determining the amount of contaminated food, and the potential place of its occurrence
- Assess the degree of risk to the population in relation to the diagnosis of disease or infection outbreak
- Genetic testing and microbiology to identify the pathogen

- Disinfection treatments in areas where there are contaminated products found
- Medical treatment of patients
- Epidemiological surveillance involving the collection of data during the epidemiological investigation, their analysis and interpretation in order of better understanding the directions of further development of the disease or infection
- Carry out hygienic measures aimed at the development and execution of post-epidemiological disposal
- Notification the appropriate services in connection with the occurrence of an increased number of cases

After carrying out planned activities to combat illness or infection, and finding that new cases no longer appear, we can proceed to further monitoring the situation. In particular, the occurrence of new cases is monitored. Upon completion of the monitoring of the situation, sanitary inspectors draw up a final report of the epidemiological investigation carried out for the outbreak of disease or infection. The following table describes the various tasks performed during the epidemiological investigation including the task role.

Each of the actions contained Figure 12.1 is in fact a separate business process. It is not possible, due to the volume of the work, to describe in detail all of these processes. However, it is worth discussing one of them. This is the most important process and on its basis the study shows the dynamic properties of business processes.

The epidemiological investigation in the event of a food–borne epidemic is part of the simulation exercise and improving procedures in the activities of the health supervision departments in cases of poisoning and infectious food–borne diseases support system. The process supports the planning and executing activities aimed at detecting the cause, sources, and mechanisms of the spread of foodborne diseases among people in each area, for a fixed population and assumptions related to the occurrence of various types, essential for the development of disease conditions. All cases of disease or deaths related to foodborne communicable diseases are recorded in the system and the information is based on the relevant paper forms supplied by the system. Each new case registered in the system results in verification of the number of cases—if it exceeds a critical number of cases or deaths defined for the disease. If the check is found to exceed the number of applications which is critical for the disease a new epidemiological investigation is launched. The basic activity of health services in the event of a food–borne epidemic, is an epidemiological investigation. The diagram of an epidemiological investigation business process realized in simulation exercise is shown in Figure 12.3.

A detailed description of the business process is as follows:

- SDSI (State District Sanitary Inspector) performs the task "Request to the governor to declare a state of emergency or epidemic". Request is completed by phone call or e-mail.

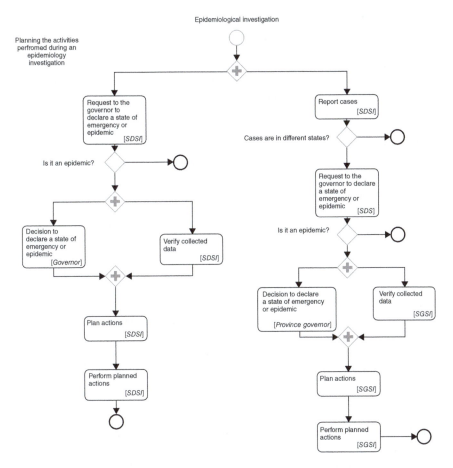

FIGURE 12.3 Business process diagram—planning the activities performed during an epidemiology investigation.

- SDSI (State District Sanitary Inspector) performs the task "Verify collected data".
- Governor performs the task "Decision to declare a state of emergency or epidemic". The operator must make a decision that will affect the control gate. The decision will involve the threat or occurrence of epidemics.
- SDSI (State District Sanitary Inspector) performs the task "Plan actions". Depending on the complexity of the situation SDSI decides to take action against epidemics or epidemic threat.
- SDSI (State District Sanitary Inspector) performs the task "Perform planned activities". It is a wide variety of activities that sanitary officer must take into action. These activities are properly documented in the document repository and records related to the epidemiology investigation.

12.3 PLANNING AND PREPARATION OF THE EVALUATION METHODS

Activities in the planning phase cover the definitions for the objectives of the computer-assisted exercises (CAX), for the preconditions to be taken care of, for the participants necessary, and finally the definition of the CAX concept. Activities in the preparation phase are the definition of the total operational situation for the exercise scenario as it should be valid at the start time of the exercise. This includes the definition of necessary data collections, the preparation of the simulation database, as well as necessary adaptations, modifications, or extensions of the simulation models involved [2]. At this stage, quantitative indicators are defined. They will be used in subsequent phases to assess the training audience. Also, the time and place of their determination is defined. The following describes the methods for assessing the results of the simulation exercises for the scenario of poisoning or infection with Salmonella. Nevertheless, these methods of exercise evaluation also apply to the assessment of different exercises, carried out under any scenario and considered any pathogen. For each case, a different set of skills will be assessed. Figure 12.1 shows the business process diagram for the simulation exercise. At the beginning of the exercises, the exercise introduction subprocess must be performed. After that the epidemic simulator is launched to begin the simulation cycle. Next the first simulation step is executed, and the system waits for trainees to analyze the situation and to send their requests (e.g., patient examination request or food production facility inspection request). The requests are processed by the system and by the exercise control team (experts). This step finishes when all the requests have been passed to execution or explained, or if the exercise step timeout has expired. After that the next simulation step is performed. The exercise control team decides if the simulation should continue. If not, the "Summary and evaluation of exercises" task is executed, the exercise evaluation team proceed to assess the results.

Subprocess "Summary and evaluation of exercises" can be modified for each exercise, in accordance with current assumptions for the assessment. An example of such a process is shown in Figure 12.5.

Each task in this process is related to the verification of specific knowledge or skills to be assessed. In the case of salmonella, assessment shall be subject to the following skills:

- Ability to evaluate the incoming information for finding an outbreak of epidemic and decide to implement the epidemiological investigation
- Ability to develop the initial report
- Ability to set a preliminary hypothesis for the diagnosis and the etiological factor
- Ability to verify data confirming the diagnosis and to actively seek such data
- Ability to develop the food intake table
- Ability to evaluate the situation of the necessity to issue restrictive, limiting, closing, or financial decisions

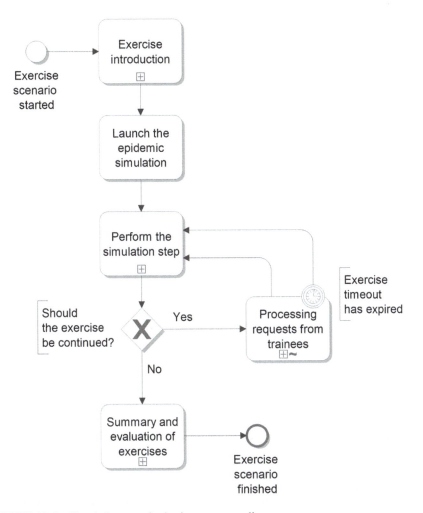

FIGURE 12.4 Simulation exercise business process diagram.

- Ability to define a confirmed or probable case
- Ability to use knowledge about the disease entity corresponding to the observed characteristics
- Ability to define a hypothesis about the source of the infection or contamination based on information from the controlled objects
- Ability to plan controls in food-distribution centers in accordance with the information held and suspicions
- Ability to compare hypotheses with the facts
- The ability to plan additional studies, analyze the results, and acquire additional data
- Ability to prepare a final report

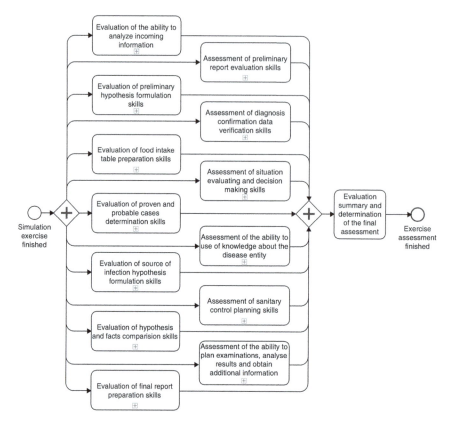

FIGURE 12.5   Sample business process for simulation exercise assessment.

At the planning stage, the method of determining the total assessment for exercise is also defined, and graphical user interfaces for assessing skills by experts are designed. The use of business processes to assess the implementation exercise gives the opportunity to design a range of input and output data for each component of the evaluation. Implementation of business processes designed in such a way can be supported by business process management systems. This gives the ability to automate process activities and allows the use of best practices regarding issuing and execution of tasks. Each task in the business process produces a component of the final grade. Not always an assessment is carried out automatically by the formula calculated based on the obtained data. In many cases it is assessed by experts. Each of them, along with the task, receives a set of data determined based on the execution history of simulation exercises. They may be calculated indicators (KPIs), charts, tabular data, various attachments, or links to external information. These data are presented using forms designed specifically for each task in the business process. These forms can also contain all the assessment data introduced by an expert. Sometimes assessment is carried out in stages. In this case, the task of assessing has to be replaced by the subprocess composed of several tasks in sequential or parallel. This approach differs significantly from conventional auto-mated or manual assessment systems.

12.4 SIMULATION EXERCISES

The design and implementation of the simulation exercise system was created to improve the procedures of sanitary inspectors during epidemics of foodborne dis-eases. In other words, the exercises are designed to better prepare inspectors to deal with potential outbreaks of foodborne poisoning or disease.

12.4.1 EMBEDDED SIMULATOR

A specialized simulator of the spread of an epidemic of foodborne diseases was constructed for simulation exercises of sanitary services. It produces a stream of information on the registration of subsequent cases of foodborne illness. Analysts of exercises and the practitioners themselves can observe and analyze the subsequent characteristics of the epidemic development shown in Figure 12.6 and declining con-taminated food as found by inspectors over time (see Figure 12.7).

12.4.2 THE METHOD OF CARRYING OUT SIMULATION EXERCISES

An important element of the organization of the simulation exercises is the simula-tor module. It is usually a discrete event-driven simulator that allows for adequate

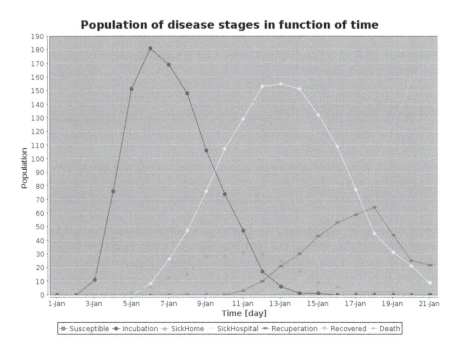

FIGURE 12.6 Population of disease stages in time.

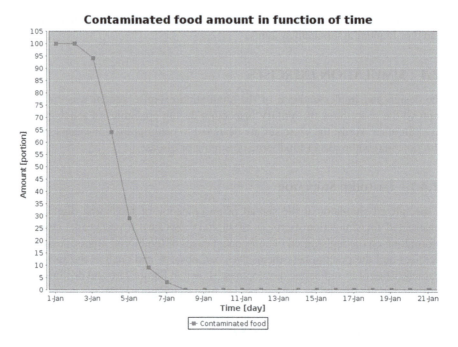

FIGURE 12.7 Contaminated food amount in time.

reflection of the environment in which the exercising teams work in an everyday manner. For activities of sanitation services in the conditions of emergence and development of food-borne disease outbreaks, the simulator has a two-fold role by implementing the following:

- Epidemic simulation for training purposes—simulator is used as a system that simulates the rise and spread of the epidemic (in the relevant processes can be adequately controlled through transmission and reception of the parameters of the processes of the organization).
- Simulation epidemic for decision support—simulator acts as a type system DSS (decision support) during the epidemic (in the relevant processes simulator provides information on past, present, and future anticipated course of the epidemic).

In the first stage of preparations, the exercise management team develops a training exercise scenario that defines the course of the simulation exercises. Based on the prepared scenario of exercises, an exercise is carried out to check the professionalism of the team. During the exercise, the exercise management team coordinates and monitors its course. The exercise management team can simultaneously coordinate many simulation exercises that are performed by people exercising. After completing the exercise, the results analysis team presents the results of the exercises for all training teams and evaluates them by showing all the good and bad elements of the recorded activities. It uses the AAR module. Each practicing sanitary inspector can evaluate his own participation in the simulation exercise.

12.4.3   THE AAR SUBSYSTEM CONCEPT

During the simulation exercises, the system supporting these exercises allows you to perform ad hoc actions, which are also recorded in the history of the actions performed. The task of the AAR subsystem is to analyze the history of all actions, regardless of their type—whether they are actions performed as part of investigative processes or ad hoc actions, however, performed in the context of a given epidemiological investigation, and to create a uniform, complete history of a given epidemiological investigation conducted as part of the exercises. The AAR subsystem also has the ability to add KPIs (Key Performance Indicator), which, by analyzing the status of processes, task execution times, and data entered by users, report the current state of the system, allowing for a more accurate evaluation of the exercise. KPIs can be presented in the form of various tabular reports, charts, or graphs (see Figure 12.6).

The collection of data for analysis is based on the evaluation indicators determined in the previous phase. History activities of individual teams, as well as data that is entered into the system, are stored in a database. This enables their analysis after exercise. The evaluation and analysis team can see the current status of the selected trained team (Figure 12.7). The preview shows the number of parameters related to execution of the simulation exercise. Evaluators can see the simulation exercise time, the registry of illness cases, map with illnesses and distribution points marked, and a graph showing the relationship between illnesses. In addition, they have access to data concerning the name of the simulation exercises and the responsible user. Cumulative summaries of many exercises performed by different training teams according to a set scenario are also available, allowing you to compare them and evaluate the progress of the participants (see Figure 12.8).

12.4.4   POST-EXERCISE ANALYSIS AND EVALUATION

Activities in the analysis phase are the review of achieved results and the replay of important situations in the scenario development related to planning and decision-making activities in the staff. Furthermore, "if–then" questions could be analyzed and answered by using the tools of the simulation system. The results of the analysis phase should be documented in a report which summarizes the "lessons learned" of the computer assisted exercises (CAX). These lessons learned should be described in such a way that they can be used for the planning of future CAX [2]. Post-exercise analysis involves the use of a variety of techniques to evaluate the data obtained during exercise. These include key performance indicators KPIs, reports, charts, and the After Action Review (AAR) technique. The presented method involves the use of each of these elements. The examples of performance indicators for simulation exercises are:

- The number of decisions of closing the distribution point—numeric value;
- The number of properly closed distribution points—numeric value;

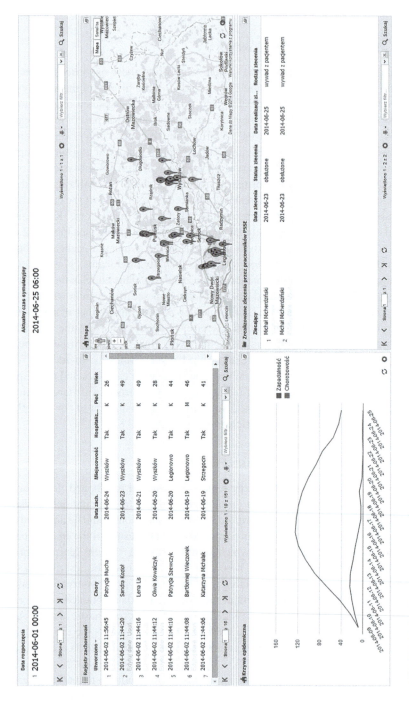

FIGURE 12.8 Key Performance Indicators presented in the form of various tabular reports, charts or graphs.

- The number of incorrectly closed distribution points—numeric value;
- The amount of tasks assigned to sanitary inspectors broken down by type (interviews with patients, patient examinations, health checks at distribution points, analysis of food samples)—tabular report;
- The epidemic curve in comparison with the best and the worst results achieved during the exercise—chart.

The After Action Review subsystem is a separate module in the simulation exercise support system (Figure 12.9). The integration with other elements of the solution is based on a shared access to databases of historical data collected from the task executions during the epidemic investigation conducted by health inspectors.

The After Action Review module uses both scenarios data and execution data of simulation exercises, also taking into account the ad hoc operations related to exercise

FIGURE 12.9   Current state of the simulation exercise.

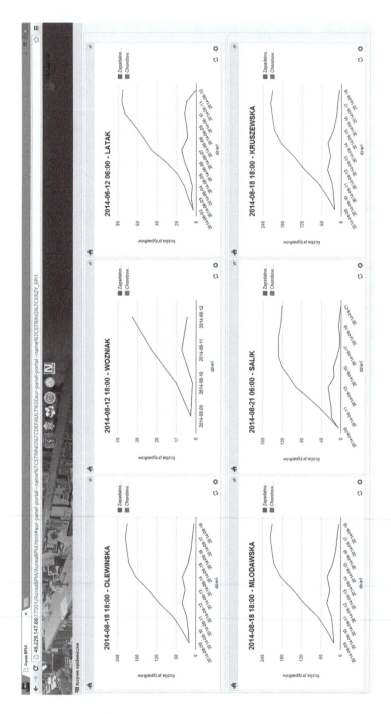

FIGURE 12.10 Cumulative summaries of many exercises performed by different training teams according to a given scenario.

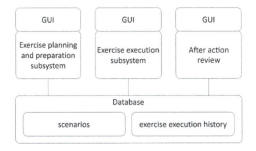

FIGURE 12.11   After Action Review subsystem environment.

audience queries and requests. The AAR module provides access to all the data of the business process, in terms of its execution events, the history of data changes, tasks' timing and cost parameters. It also guarantees access to aggregated statistical data of other processes, which stem from previously conducted similar simulation exercises. It allows to define and calculate KPIs, as well as define custom reports based on historical process data. It also provides comprehensive data visualization by means of summary tables and charts and gives the opportunity to view the history of simulation exercises on the timeline as similar to a "filmstrip". Additionally, it provides synchronization functionality for KPIs presenting the current state of the process (the frame of the filmstrip) and tracks changes in the process data, including the ability to perform "a journey backward in time" and boot process from a designated time in the past. Based on the results obtained during two simulation exercises held in Warsaw last year, significant differences in the pace and accuracy of execution of tasks during the epidemiological investigation could be observed. In these exercises, six teams of sanitary inspectors practiced simultaneously. Their task was to conduct epidemiological investigations in the case of a foodborne outbreak.

12.5   EXPERIENCES AFTER CONDUCTING THE SIMULATION EXERCISES

During each simulation step, the time shift of the corresponding interval is also simulated. Using computer simulation in the spread of the epidemic, it is possible to simulate the entire course of the epidemic in a short time. In mature simulation exercises, the AAR mechanism allows us to view and analyze the course of exercises by teams performing exercises. The analysis module with AAR at its disposal will be able to track the history of activities during the simulation step by step. The conclusions of this analysis will be used to evaluate and improve procedures and actions to combat epidemics of food poisoning and diseases. The technology used to visualize the results of the simulation exercises performed in the After Action Review module provides an overview of the process simulation history also during the exercise. Individual results of simulation experiments are stored in a database constructed, and at the user's request simulation process is restored to any moment of an outbreak of the disease. Depending on the needs, the simulation process is repeated. Simulation

FIGURE 12.12   Photo of the person taking part in simulation exercises and post-simulation analysis.

exercises consist of the creation of an artificial epidemic and the examination of participants' actions in preventing the development of epidemics. You can also request for the comeback of conducting exercises at any moment of the epidemic, and after changing some parameters of the epidemic, simulation experiment for the revised terms of development and the prevention of epidemics sanitary service activities and decisions of the authority of these services may be carried out. Any results obtained in this way are stored in a database and can be used to illustrate the course of the epidemic of foodborne illness. At the same time, these results become the basis for assessing the members of the teams participating in simulation exercises. The simulation exercise system was positively assessed by the sanitary services participating in the exercises.

A large series of simulation exercises were conducted for sanitary inspectors (see Figure 12.12). Various exercise scenarios were used. Each simulation exercise lasted approximately 3–4 hours.

12.6   CONCLUSIONS

A programming environment was launched and tested on the foodborne disease epidemic development model and the work of sanitary inspectors during the development and duration of the epidemic. The results of solved research problems that appeared in connection with the creation of a simulation environment for the organization of simulation experiments have been published (Nowicki, 2012; Nowicki et al., 2014). Simulation exercises are one of the most important elements of consolidating good practices for sanitary inspectors. The organization of simulation exercises for sanitary inspections requires many years of preparation of both training teams and the team preparing the exercises. User interfaces will be characteristic of the natural working environment of sanitary inspectors. Internal interfaces between the components of the simulation exercise support system ensure good communication, archiving, situational visualization, and multi-variant analysis of the course of exercises.

ACKNOWLEDGMENT

This work was supported in part by the grant UGB 22–799 financed by Military University of Technology, Warsaw, Poland.

REFERENCES

Cayirci E., Marincic D. (2009). *Computer Assisted Exercises and Training: A Reference Guide*, John Wiley and Sons.

IEEE Std 1516 TM. (2010). *IEEE Standard for Modeling and Simulation (M&S) High Level Architecture (HLA)—Framework and Rules*. IEEE.

Jasiulewicz-Kaczmarek M. et al. (2018). Implementing BPMN in Maintenance Process Modeling, *Advances in Intelligent Systems and Computing*, Volume 656, DOI: 10.1007/978-3-319-67229-8_27.

Lateef F. (2010). Simulation-Based Learning: Just Like the Real Thing, *Journal of Emergencies, Trauma, and Shock*, Volume 3, No. 4, pp. 348–352.

Nowicki T. (2012). The Method for Solving Sanitary Inspector's Logistic Problem, in: *Production Management—Contemporary Approaches—Selected Aspects*, Publishing House of Poznan University of Technology, pp. 23–34, ISBN: 978-83-7775-189-3.

Nowicki T., Pytlak R., Waszkowski R., Bertrandt J., Kłos A. (2014). *Formal Models of Sanitary Inspections Teams Activities*, International Conference on Food Security and Nutrition, Johannesburg, South Africa, February 10–11.

Nowicki T. et al. (2016). Data Flow Between RFID Devices in a Modern Restricted Access Administrative Office, *MATEC Web of Conferences*, Volume 76, DOI: 10.1051/matecconf/20167604004.

Schirlitzki H. J. ed. (2007). *Exercises—Planning and Execution in Integration of Modelling and Simulation*, NATO Research and Technology Organization, ISBN 978-92-837-0046-3.

Waszkowski R. (2018). Multidimensional Modeling and Analysis of Business Processes, *IOP Conference Series-Materials Science and Engineering*, Volume 400, DOI: 10.1088/1757-899X/400/6/062031.

Waszkowski R., Kiedrowicz M. (2015). Business Rules Automation Standards in Business Process Management Systems, in: *Information Management in Practice*, Faculty of Management, University of Gdańsk.

13 MIG Welding Quality Improvement Study for Joined St37 Material Using Building Constructions

Yusuf Tansel İç, F. Soner Alıcı, Atahan Erdağ, Nehir Atila, and İrem Sayın

CONTENTS

13.1 INTRODUCTION

The use of MIG-welded steel rods in construction projects has become widespread, and it has become a necessity for welded joints to have the desired quality standards. These standards must be complied with for the buildings to correspond to earthquake codes. St37 steel is the most common type used for reinforced concrete structural members. In construction projects, welding can be preferred for reinforcing bars' splicing according to the reinforcing details of structural members. Therefore, for this type of splicing, welding quality and its strength become important. In the light of these, strength analyses of the welded joints of St37 rods are investigated in this

DOI: 10.1201/9781003383444-13

chapter, and appropriate welding factors (parameters) that provide the highest joint strength are determined. However, the welding process quality should be improved with appropriate welding factor values crucial for RC member strength. Structural problems may occur in welded joints of RC members that are not performed with appropriate factor values. This situation can adversely affect the safety of buildings and their strength against natural disasters such as earthquakes.

The strength properties of the weldable St37 steels in the current TS 708 standard have been determined according to the Turkish Earthquake Code 2018 (TEC 2018). On the other hand, welding conditions for welded joints are given in ISO 9692–1:2013, related to the topic of weld design with suitable electrodes. In this code, there are no specific values for the design factors and the ranges in which the relevant values can be used in the welded structures. Therefore, it is expected that welded joints made with different factor values have different strength properties.

In the literature, some studies are using the MIG/MAG welding process for St37 material. For example, Utkarsh et al. [1] investigated the Gas Metal Arc Welding (GMAW) process that shows the effect of current, voltage, gas flow rate (l/min), and speed (m/min) on the ultimate tensile strength of St37 low alloy steel material using the Taguchi's L9 orthogonal array. Muzakki et al. [2] analyzed the welding parameters of Metal Inert Gas (MIG), which affected the tensile strength of the St37 weld joint. Mukhraiya et al. [3] examined the interaction between process parameters and torsional rigidity of the St37 welded rods. Ebrahimnia et al. [4] investigated the influence of variation in the shielding gas composition on the weld properties of the steel St 37–2. Patel and Patel [5] optimized GMAW factors of current, voltage, gas flow rate (l/min), and weld depth of St37 low carbon alloy steel material using the Taguchi methods. Adin and İşcan [6] proposed an optimization study related to getting better mechanical properties of St37 medium carbon steel joints joined by the MIG welding process using the Taguchi method. Karami et al. [7] compared the tungsten Inert Gas (TIG) and friction stir welding methods based on the formability and mechanical properties of the welded St37 sheets. Hamzayw et al. [8] proposed optimization of the GMAW process parameters using Taguchi's optimization approach. They presented the optimum GMAW process parameters considering the welding voltage, wire diameter, wire feed rate, and the CO_2 gas flow rate that achieve the best ultimate tensile strength of St37 welded joints. Ampaiboon et al. [9] analyzed the effect of welding parameters on the ultimate tensile strength of structural steel, St37–2, welded by metal active gas (MAG) welding. They used a fractional factorial design for obtaining the signature of six parameters: wire feed rate, welding voltage, welding speed, travel angle, tip-to-work distance, and shielded gas flow rate.

Researchers used different parameters affecting welding process for optimization studies in the literature. The majority of the studies focused on the interpretation and analysis of the results obtained. Similarly, in the analysis section of this study, after the experiments and tests are over, the linear programming (LP) model is proposed. This specification is the main contribution of the methodological perspective. Also, a quality standard integrated with experimental design methodology was proposed in this chapter. After all, a new approach for the St37 MIG welding process optimization was developed.

13.2   MATERIALS AND METHODS

St37 is one of the seven types of steel used in general building materials. St37 steel is produced by the cold drawing process (transmission) in which the steel is reprocessed by the hot production. As a result of the cold drawing technique, the steel gains new properties, and becomes more durable. As a result of this process, the material characteristics of St37 steel changes. These changes can be listed as:

- The steel comes to precise measurement tolerances.
- The steel has a higher level of surface quality than hot-rolled.
- The tensile quality of the product is increased the hardness of the product increases.
- The creep state decreases.

The chemical properties and mechanical properties of St37 material are given in Tables 13.1 and 13.2 [10].

13.3   REINFORCED CONCRETE STEEL STANDARD

The standard of reinforced concrete steel bars is TS 708 (in Turkish). Steels defined in the old regulation TS 708:1996 were canceled, and new steels were defined in April 2010. TS 500:2000 Article 3.2 states: *The steels to be used as concrete reinforcement must comply with TS 708.* Symbols of reinforced concrete steel are defined in S according to the TS 708:2010. In the code, steels are labeled as S 220, S 420, B 420B, B 420C, B 500A, B 500B, and B 500C. Reinforced concrete steel classes and mechanical properties are given in Table 13.3 with respect to TS708:2010.

According to TEC 2018, the ribbed steel bars should be used for the design of reinforced concrete structural members (beams, columns, and shear walls) to provide proper bonding between steel and concrete, and to correspond to the design requirements stated in the code for ductile behavior. They have maximum yield strength of

---

TABLE 13.1
Chemical Properties of St37 [10]

Material				Chemical Properties (% Weight)						
St37	C	Si	Mn	P	S	Cr	Ni	Mo	W	Fe
	0.117	0.125	0.295	0.025	0.030	0.087	0.25	0.007	0.003	Ball

TABLE 13.2
Mechanical Properties of St37 [11]

Material	Tensile Strength (N/mm^2)	Yield (N/mm^2)
St37	360–470	225–235

420 N/mm², the ratio of minimum experimental elongation to experimental yield equal to 1.15, the ratio of maximum experimental yield to characteristic yield equal to 1.3, and minimum elongation at break equal to 10%. Accordingly, it can be said that the B420C is the most suitable steel that can meet these conditions. So, this steel material is used in the following sections of this study.

13.4 WELDING PROPERTIES

We used a gas metal arc welding (MIG) process to join B420C steel rods in this study. MIG welding is an electric arc welding between a continuously fed welding wire and the workpiece by gas metal arc welding with a melting electrode. In MIG welding, a filler material or welding wire ignites the arc as it contacts the workpiece. It is used as a melted wire material addition. Shielding gas mentioned previously also flows into the gas nozzle to protect the melted point against the reactive oxygen. This gas mixes with oxygen during welding, and thus prevents oxidation in the arc and in the weld pool [12].

The required temperature in gas arc welding is generated by the arc formed between a continuously fed and melting wire electrode and the weld pool by the resistance heating of the electrode through which the welding current passes. The electrode is a bare wire and is conveyed to the welding area at a constant speed by an electrode-feeding device. The bare electrode, weld pool, arc, and areas of the base metal adjacent to the weld area are protected against atmospheric pollution by a suitable gas or gas mixture supplied from outside and delivered to the area through a gas nozzle.

13.5 WELDING STANDARD OF ST37 STEEL

The standards for welding steels are listed next [13]:

- TS EN ISO 2560: 2009—Welding Consumables—Mailed Electrodes for Manual Metal Arc Welding of Non-Alloy and Fine-grain Steels—Classification

TABLE 13.3
Reinforced Concrete Steel Classes and Mechanical Properties

Properties	Class						
	S220	S420	B 420B	B 420C	B500 B	B 500C	B 500A
Yield strength, N/mm^2	≥220	≥420	≥420	≥420	≥500	≥500	≥500
Tensile strength N/mm^2	≥340	≥550	N/A	N/A	N/A	N/A	≥550
Tensile strength/yield strength	≥1.2	≥1.15	≥1.08	≥1.15 and <1.35	≥1.08	≥1.15 and <1.35	N/A
Experimental yield strength/characteristics yield strength	N/A	≤1.3	N/A	≤1.3	N/A	≤1.3	N/A
Unit breaking elongation	≥18	≥10	≥12	≥12	≥12	≥12	≥5
Maximum elongation (%)	N/A	N/A	≥5	≥7.5	≥5	≥7.5	≥2.5

TABLE 13.4

Welding Joints According to TS EN ISO 9692–1:2013 Standard [14]

Ref.	Material Thickness t (mm)	Symbol (In Accordance with ISO 2553[17])	Cross Section	Angle	Gap b (mm)	Thickness of Root Face c (mm)
1.3	$3 < t \leq 12$		α	$40 \leq \alpha \leq 60$	≤ 4	≤ 2
1.9.2	$3 < t \leq 10$		β	$35 \leq \beta \leq 60$	$2 \leq b \leq 4$	$1 \leq c \leq 2$

- TS EN ISO 9692–1:2013—Welding and Similar Processes—Recommendations for Welding Button Preparation—Part 1: Manual Metal Arc Welding of Steels, Gas Protected Metal Arc Welding, Gas Welding, TIG Welding, and Batch Welding.

In Table 13.4, appropriate welding preparation standards are presented according to the TSENISO9692–1:2013 standard [14].

13.6 EXPERIMENTAL DESIGN

The experimental design was used as a method in this study. A design in which all possible combinations of factor levels are explored is called factorial design [15]. In this study, there are three factors, each with two levels, and the fractional factorial design method is applied to reduce the number of experiments and, a total of eight experiments are carried out as a result of $2^{3-1} = 4$ experiments with two replications.

In the study, factors are named as A, B, and C for the estimation of the main effect in factorial experiments. The levels of A, B, and C factors in the experiment are called with "Low" and "High" and with the expressions of "−1" and "+1". These factors with the levels can be listed as A high (+1), A low (−1), B high (+1), B low (−1), C high (+1), and C low (−1), and are shown in Table 13.5. Response values can be obtained against the values given in this table from the tensile strength test results (Table 13.6). The welding process and test conditions have uncontrollable factors such as temperature, humidity, operator efficiency, machine efficiency, and so on. So, we used Taguchi's S/N formulation for the robustness of the test results [16].

TABLE 13.5

Experimental Design Factors and Their Levels

Symbol	Factors	Level	
		−1	+1
A	Gap	2 mm	4 mm
B	Current	70 A	120 A
C	Cross section	$\bigvee$	$\bigvee$

13.7 APPLICATION

The rod diameter was assumed to be constant and taken as equal to 10 mm. According to TS708 welding standards, it was deemed appropriate to use two types of weld angles in our experiment. The test rods with a length of 400 mm should be divided into two equal parts of 200 mm in length. The welding rod preparation process was carried out using the saw machine in the Baskent University Engineering Faculty Machine Tools laboratory (Figure 13.1). Then, the appropriate rods for the welding process were obtained according to the experimental design (Figure 13.1-a-d).

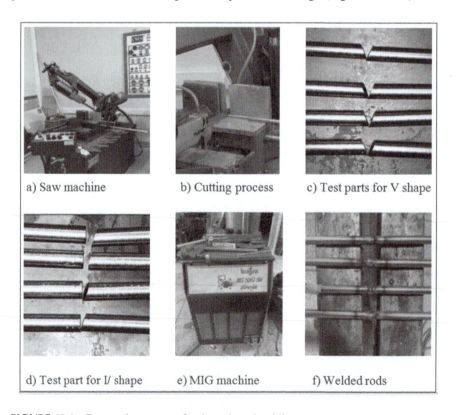

a) Saw machine   b) Cutting process   c) Test parts for V shape

d) Test part for I/ shape   e) MIG machine   f) Welded rods

FIGURE 13.1 Preparation process for the rods and welding process.

13.7.1 PERFORMING THE WELDING PROCESS

The MIG welding process in this study was carried out at the Başkent University sheet metal facility (Figure 13.1-e, f). A total of eight welding operations were implemented with four experiments and two replications according to the 2^{3-1} experimental design. Welded B420C steel rods are shown in Figure 13.2.

13.7.2 PERFORMING THE TENSILE TEST

The tensile test was processed in the Gazi University Vocational School Laboratory and the specimens were subjected to tensile tests in the computerized tensile test machine (Figure 13.2). The tensile speed in the experiment is 5 mm/min. So, eight different welded rods are tested following the experimental design, and test results are presented in Figure 13.3 and Table 13.6. As a result, some of the rods broke off from the main material region others broke from the heat-affected zone (HAZ). The HAZ is the region that is under the influence of heat, since the rising temperature during the welding process affects the internal structure, and thus material properties. The images from the test results are shown in Figure 13.2.

13.7.3 ANOVA RESULTS

After collecting data about the tensile strength of the welded rods, we analyzed the experimental results using the ANOVA analysis in the MINITAB program. The obtained results are shown in Table 13.7 and Table 13.8. It is realized in Figure 13.3 that the data form an approximately straight line along the line. So, the normal probability plot of the residuals is a nearly linear line supporting the condition that the error terms are normally distributed. To discuss the relevance of the normality assumption, we looked over the Gauss Markov theorem, which indicates that the ideal linear regression estimates are both unbiased and have the least amount of variance, a property called the "best linear unbiased estimators" (BLUE) [18]. The linear

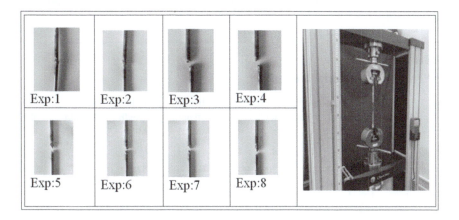

FIGURE 13.2   Pictures of tensile test results (left) and tensile test equipment (right).

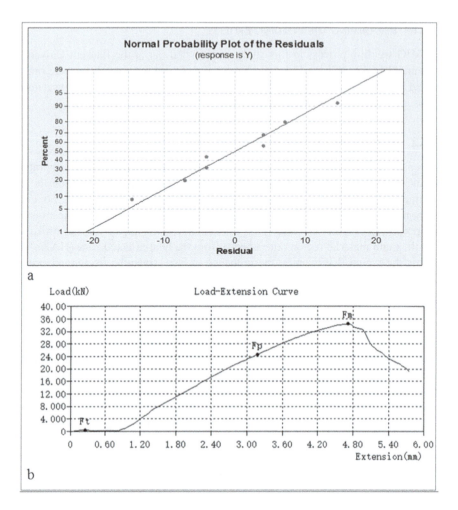

FIGURE 13.3   (a) Normal probability plot of the residuals. (b) Confirmation test's load-extension curve.

TABLE 13.6
2^{3-1} Fractional Factorial Design

Exp.	Experimental Design: Coded Units			Experimental Design: Uncoded Units			Tensile Strength (MPa) R_m	Yield Strength (MPa) R_p	Tensile Strength/ Yield Strength	S/N for R_m[a]
	A	B	C	A	B	C				
1	1	1	1	4	120		442	382	1.157	52.91
2	-1	1	-1	2	120		458	379	1.208	53.22
3	1	-1	-1	4	70		360	217	1.687	51.13

Exp.	Experimental Design: Coded Units			Experimental Design: Uncoded Units			Tensile Strength (MPa) R_m	Yield Strength (MPa) R_p	Tensile Strength/ Yield Strength	S/N for R_m[a]
	A	B	C	A	B	C				
4	1	-1	-1	4	70	∨	352	351	1.003	50.93
5	-1	-1	1	2	70	∨	497	434	1.145	53.93
6	1	1	1	4	120	∨	471	424	1.111	53.46
7	-1	-1	1	2	70	∨	505	458	1.103	54.07
8	-1	1	-1	2	120	∨	450	369	1.220	53.06
Average							442.625	376.75	1.102	52.84

[a] $S/N = -10\log (1/R_m^2)$

TABLE 13.7
Estimated Effects and Coefficients for Y (Coded Units)

Term	Effect	Coef	SE Coef	T	P
	Constant	442.63	4.267	103.74	0.000
A	-69.75	-34.88	4.267	-8.17	0.001
B	25.25	12.62	4.267	2.96	0.042
C	72.25	36.13	4.267	8.47	0.001

S = 12.0675 R–Sq = 97.36% R–Sq (adj) = 95.37%

TABLE 13.8
Analysis of Variance for Y (Coded Units)

Source	DF	Seq SS	Adj SS	Adj MS	F	P
Main effects	3	21445.4	21445.4	7148.5	49.09	0.001
Residual error	4	582.5	582.5	145.6		
Pure error	4	582.5	582.5	145.6		
Total	7	22027.9				

regression equation predicts BLUE when the errors have a mean of zero, are uncorrelated, and have equal variance across different values of the independent variables [18]. The normality assumption is thus unnecessarily to get estimates with the BLUE property. However, in small sample size settings, the standard error estimates may be biased (and hence confidence intervals and P-values, as well) when the errors do not follow a normal distribution [18].

According to the results, all factors have a significant effect on tensile strength. Now, a linear programming model can be obtained using Table 13.7.

$$\text{Max Y} = 442.63 - 34.88\,A + 12.62\,B + 36.13\,C \qquad (13.1)$$

TABLE 13.9

Confirmation Test Results

Performance criteria	Non-welded Material	Average Values for Experimental Design Results	Confirmation Test		
			Results	Performance for Non-welded Material (%)	Performance for Experimental Design Average Values (%)
Tensile strength (MPa) R_m	538	442.63	439	−18.40	−0.82
S/N for $R_m{}^a$	54.62	52.84	52.85	−3.24	0.02
Yield strength (MPa) R_p	488	376.75	313	−35.86	−16.92
Tensile strength/ yield strength	1.102	1.204	1.403	27.27	16.49

Subject to:

$$1 \le A \le +1, \tag{13.2}$$

$$1 \le B \le +1, \tag{13.3}$$

$$1 \le C \le +1. \tag{13.4}$$

We used the MS Excel Solver tool to solve the aforementioned LP problem and we get optimal factor levels as follows: A = −1 (b = 2 mm), B = 1 (120 ampere), C = 1 (I/). After getting optimal factor levels, we made a confirmation test using the three samples and reached the new tensile strength values shown in Table 13.9. According to Table 13.9, the results obtained are similar with the predicted tensile strength values. This test indicates that the presented model is suitable for the St37 welding process parameter estimations which are crucial for the building structure's welding process.

13.8   CONCLUSION

We presented an experimental design study to determine the welding parameter values that optimized the welding quality in MIG welding of St37 (B420C) steel. Based on the results obtained from the proposed study, the following results can be concluded: (i) The factor combinations and their levels for optimum tensile strength and, therefore, for optimum strength of the St37 material under the constant load are A: −1, B:1, C:1 (i.e., gap 2 mm, weld current-120 ampere, and section I/). (ii) The contribution of the proposed study for design to the quality characteristic seems to be minimal. But, the welding geometry design perspective is important to practitioners. (iii) We reached a competitive tensile strength performance considering S/N values comparisons. (iv) Major contribution of the proposed study is the tensile strength/ yield strength ratio. We reached the best performance for this ratio which is crucial for construction safety (Figure 13.3-b).

ACKNOWLEDGMENTS

The authors would like to thank Gazi University Faculty of Technology Faculty Members Prof. Dr. Adem Acır and Gazi University Technical Sciences Vocational School member Dr. Ramazan Çakıroğlu for facilitating the tensile test experiments.

DATA AVAILABILITY STATEMENT

The authors confirm that the data supporting the findings of this study are available within the book/chapter.

REFERENCES

1. Utkarsh, S., Neel, P., Mahajan, M. T., Jignesh, P., Prajapati, R. B.: Experimental investigation of MIG welding for ST-37 using design of experiment. *International Journal of Scientific and Research Publications*, 4(5), 1–5 (2014).
2. Muzakki, H., Prasetyo, T., Umam, M. S. U., Lumintu, I., Hartanto, D.: Effect of metal inert gas welding process parameters to tensile strength on ST 37 steel sheet joint. *Journal of Physics: Conference Series IOP Publishing*, 1569(3), 032055 (2020).
3. Mukhraiya, V., Yadav, R. K., Jathar, S.: Parametric optimisation of MIG welding process with the help of Taguchi method. *International Journal of Engineering Research & Technology (IJERT)*, 3(1) (2014).
4. Ebrahimnia, M., Goodarzi, M., Nouri, M., Sheikhi, M.: Study of the effect of shielding gas composition on the mechanical weld properties of steel ST 37–2 in gas metal arc welding. *Materials & Design*, 30(9), 3891–3895 (2009).
5. Patel, T., Patel, S. C.: The effect of process parameter on weld depth in GMA welding process. *International Journal for Innovative Research in Science & Technology*, 1(11) (2015).
6. Adin, M. Ş., İşcan, B.: Optimization of process parameters of medium carbon steel joints joined by MIG welding using Taguchi method. *European Mechanical Science*, 6(1), 17–26 (2022).
7. Karami, V., Dariani, B. M., Hashemi, R.: Investigation of forming limit curves and mechanical properties of 316 stainless steel/St37 steel tailor-welded blanks produced by tungsten inert gas and friction stir welding method. *CIRP Journal of Manufacturing Science and Technology*, 32, 437–446 (2021).
8. Hamzawy, N., Zayan, S. A., Mahmoud, T. S., Gomaa, A. H.: *On the optimization of gas metal arc welding process parameters*, 1–7. https://feng.stafpu.bu.edu.eg/Mechanical%20Engineering/5817/publications/nadia%20hamzawy%20mohamed_Nadia%20Paper.pdf.
9. Ampaiboon, A., Lasunon, O. U., Bubphachot, B.: Optimization and prediction of ultimate tensile strength in metal active gas welding. *The Scientific World Journal* (2015). doi: 10.1155/2015/831912.
10. Yazdani, M., Toroghinejad, M. R., Hashemi, S. M.: Investigation of microstructure and mechanical properties of St37 steel-Ck60 steel joints by explosive cladding. *Journal of Materials Engineering and Performance*, 24(10), 4032–4043 (2015).
11. www.theworldmaterial.com/1-0037-material-st37-steel-din-17100/.
12. Groover, M. P.: *Fundamentals of modern manufacturing: Materials, processes, and systems*. John Wiley & Sons (2020).
13. TS EN ISO 2560:2009: *Welding consumables—mailed electrodes for manual metal arc welding of non-alloy and fine-grain steels—classification* (2009). https://www.iso.org/standard/45947.html.

14. TS EN ISO 9692–1:2013: *Welding and similar processes—recommendations for welding button preparation—part 1: Manual metal arc welding of steels, gas protected metal arc welding, gas welding, tig welding, and batch welding* (2013). https://www.iso.org/obp/ui/#!iso:std:62520:en.

15. Montgomery, D. C.: *Design and analysis of experiments*. John Wiley & Sons (2017).

16. Phadke, M. S.: *Quality Engineering Using Robust Design*. Prentice Hall (1989).

17. EN ISO 2553:2019: *Welding and allied processes—symbolic representation on drawings—welded joints*. https://www.techstreet.com/standards/iso-2553-2019?product_id=2039632.

18. Schmidt, A. F., Finan, C.: Linear regression and the normality assumption. *Journal of Clinical Epidemiology*, 98, 146–151 (2018).

14 Control of the Process of Plasma-Arc Spraying by the Method of Control of Dynamic Parameters of Condensed Phase Particles

Oksana Isaeva, Marina Boronenko, and Pavel Gulyaev

CONTENTS

14.1   INTRODUCTION

Recently, much attention has been paid to various processes using low-temperature plasma, and in particular to the process of plasma spraying. Despite the fact that the plasma spraying method was created relatively long ago, there are several unresolved issues in this area that are related to the choice of optimal spraying modes (Khafizov et al., 2014).

The main tasks of the diagnostic tools used are the measurement of variable particles in the spray stream, that is, velocity, temperature, flow, trajectory, and size distribution, which affect the microstructure and properties of the sprayed coatings (Zhang & Sampath, 2009). Basically, the measurements are based on two-wave pyrometry and the time-of-flight method. Examples of thermal spray sensors are Tecnar DPV2000, Oseir SprayWatch, and Tecnar Accuraspray. In this case, the parameters of individual particles (DPV2000) (Mauer et al., 2007) or their ensemble (group of particles) (SprayWatch, Accuraspray) (Hamalainen et al., 2000) are monitored. However, building a closed loop process control system that controls particle performance remains a long-term goal because knowledge of the relationship between particle performance and coating properties is incomplete (Mauer et al., 2011).

An enthalpy probe, optical emission spectroscopy, and computed tomography are used to diagnose plasma flow. The enthalpy probe is a probe for studying the enthalpy, temperature, and velocity fields of hot and plasma gas flow. Accuracy within 1% can be achieved in two to three calibrations. This method has a limited ability to withstand high thermal loads, so it can only be used at lower plasma power; or at large distances from the torch optical emission spectroscopy is a non-contact method, which allows measurements near the nozzle. However, under atmospheric conditions, this method is hampered by broadening effects and continuous background radiation. The reliability of the results is given only under local conditions of thermal equilibrium. Computed tomography makes it possible to spectroscopically measure the plasma temperature even of non-rotationally symmetric jets, since it does not depend on the conditions of axial symmetry. However, this method requires sophisticated software to reconstruct the temperature distribution (Mauer et al., 2011). Methods for controlling the parameters of the deposited material can be divided into single-particle and ensemble methods. Single-particle methods estimate the temperature of individual particles, which makes it possible to establish a more complete relationship between particle characteristics and coating properties. In this case, the relative error of temperature measurement will be about 5% (Fauchais & Vardelle, 2010). If the process is non-stationary, the average value obtained experimentally will be very different from the true one. Single-particle-based sensors (DPV2000, TECNAR Automation Ltd (Mauer et al., 2007), Flux Sentinel (Cyber Materials LLC) (Wroblewski et al., 2010), are more complex and time-consuming than ensemble methods (SprayWatch (Oseir Ltd.) (Hamalainen et al., 2000) and Accuraspray (TECNAR Automation Ltd.) (Mauer et al., 2007). The SprayWatch has the additional benefit of equipping the CCD with an electronic shutter, which enables digital imaging.

Particle temperature measurements are based on the measurement of radiation from hot, incandescent particles in two or more wavelengths or colors (Fincke et al., 2001). Such measurements are carried out using the method of multicolor pyrometry. Ensemble mean particle temperature at 10 Hz is measured with an Air Particle Pyrometer (IPP) probe (Inflight Ltd. Co.) (Swank et al., 1995), ensemble velocities, and particle sizes, ThermaViz system (Stratonics Inc.) (Craig et al., 2003). Changes in the emissivity of sprayed materials with temperature increase the error up to 25% (Salhi et al., 2005).

The optimization of the sputtering process is used to improve the characteristics of the resulting coating, monitor the instability of the process, and check the reproducibility. In general, to control the deposition process, it is necessary to build a model and optimize the process (Chen, Shuang Hchuan, et al., 2014). To select the optimal conditions, a comprehensive analysis of the interaction of the particles of the sprayed material with the plasma flow is necessary, since the process parameters are related to the characteristics of the coating. Therefore, an understanding of the relationships between the process, microstructure, and properties of the resulting coating is necessary (Sampath et al., 2009). Usually, the deposition process is developed by setting up a limited set of technological experiments on specific equipment using a powder material of one or another fractional composition, followed by a study of the characteristics of the deposition coatings (Table 14.1) (Solonenko et al., 2011b).

TABLE 14.1
Basic Methods for Optimizing the Deposition Process

Principle/ Optimization Method	Measured Parameters	What Makes It Possible to Achieve	Restrictions
Obtaining the maximum utilization of the material and the minimum porosity of the coating	Coating porosity, powder consumption, adhesion strength, hardness	Improving the antifriction characteristics of ceramic wear-resistant plasma coatings under conditions of high-temperature corrosion	Only for materials Al_2O_3-TiO_2-12% (MoS_2-Ni), Al_2O_3-TiO_2-12% (CaF_2-Ni) (Okovity & Panteleenko, 2015)
Application of artificial neural networks	Arc current; spray distance; dispersion of titanium powder. Similar parameters for hydroxyapatite coatings	Obtaining maximum adhesion of the coating to the base and improving its porosity	Only titanium and hydroxyapatite powders (Timofeev et al., 2021)
Use of a single experimental plan	Flow rate of argon and hydrogen; deposition efficiency, porosity, microhardness	Revealing the influence of the studied process parameters on the deposition efficiency, porosity, and microhardness	Only for yttria-stabilized zirconia coatings (Li et al., 2005)
Investigation of the influence of certain parameters on the microstructure using statistical methods	Powder size, spray nozzle size, total consumption of plasma gas, $H_2 + N_2$ ratio, excess of the total gas consumption	The results of the microstructure evaluation are used to generate regression equations to predict the microstructure of the coating based on the process parameters	For NiCrAlI coatings deposited with the Mettech Axial III system (Gao et al., 2012)
Application of the fuzzy order preference method by similarity with the hybrid ideal solution method	Primary gas flow rate, arc current	Proof of the importance of the primary gas flow rate in obtaining a coating with better properties	For atmospheric plasma spraying (Swain et al., 2021)
The concept of "technological maps"	Temperature and particle velocity	Improved deposition efficiency (DE) as well as process efficiency; reduction in coating production costs	To study the fundamental relationship of process, structure, and properties in high-velocity oxygen fuel (HVOF) sprayed coatings (Valarezo et al., 2019)

Thus, the optimization problem is usually solved individually for a specific equipment for a specific case, by searching for the technological mode of the installation that provides the required characteristics of the coating, and maintaining it at the stage of operation.

However, the quality of the coating does not directly depend on the technological mode itself, but on the parameters (velocity and temperature) of the sprayed particles before impact with the surface of the part (Ermakov et al., 2014; Solonenko & Gulyaev, 2009; Gulyaev & Solonenko, 2013).

Therefore, the control of the spraying process can be carried out not according to the regime parameters, but according to the velocity and temperature of the particles of the sprayed material. Sensors for measuring the temperature and velocity of the particles of the powder in the atomized stream can be used to control T and V in the implementation of feedback.

To determine these characteristics, only non-contact methods are suitable, among which optical methods have the highest resolution (Gulyaev, 2012, Solonenko et al., 2011).

An expert system was proposed (Liu et al., 2013) to optimize the coating deposition process and based on an artificial neural network (ANN) and a fuzzy logic controller (FLC). The temperature measurement error is 3% and speed 2%. This expert system now uses a DPV 2000 sensor (TECNAR Automation, St-Bruno, QC, Canada), which requires real-time monitoring at a high sampling rate.

Modern high-speed cameras and digital methods for processing their signal make it possible to solve the problem of measuring the velocity of the condensed phase with an error of 1–2% (Boronenko, 2014). The error in measuring the velocity of individual particles is ±1%. The methodological error in measuring the particle temperature, associated with the spectral dependence of the emissivity of the material, for the accepted experimental conditions is in the range from + 60 to + 100 K, and the random measurement error was ±2% (Kuzmin, 2020).

One of the modern methods of noncontact temperature measurement is the method of spectral-brightness pyrometry (SBP) (Gulyaev I.P. et al., 2017). SNP combines the best qualities of brightness and spectral pyrometry in measuring instruments: high-temperature sensitivity and accuracy in determining the reference (single) temperature. The most significant property of the SNP is the self-calibration of the measuring instrument in the in situ mode, i.e., the calibration is carried out according to the object of study in the process of observing it. It allows you to abandon the calibration procedure using special temperature standards or, conversely, expand the range of materials and processes for calibrating secondary pyrometers by comparing their readings with instrument data based on the SNP method (Gulyaev I.P., 2017, 2018a). The SNP approach is implemented in the YuNA (Yugra—Novosibirsk—Altai) thermal imaging system, aimed at determining the properties of the condensed phase of the GTN flows and additive synthesis (Gulyaev I.P., 2018b). The temperature measurement error was less than 1% (Dolmatov, 2020).

Thus, the real-time control of the deposition process requires building a model and optimizing the process. But the quality of the coating directly depends on the speed and temperature of the sprayed particles before impact with the surface of the part. Therefore, the main models to be built are models of heating and acceleration of particles in a plasma flow. Statistical particle temperature density distributions determined using single-particle methods provide insight into the melting state of the particles, which is a key characteristic for obtaining specific coating characteristics.

The purpose of this study is to test the YuNA thermal imaging system method for controlling the process of plasma-arc spraying. The issue of changing the functionality of the operator who controls the deposition process, due to the digitalization of technology, is also considered.

14.2 THEORETICAL MODEL OF MOTION AND HEATING OF PARTICLES IN PLASMA

The theoretical model of the motion and heating of macroscopic particles in a plasma flow is based on the classical principles of heat and mass transfer in heterogeneous plasma flows, which in each specific case can be applied with allowance for the processes of melting, sublimation, and convective mixing and the formation of gas cavities in the substance of molten particles (Boronenko et al., 2012; Gulyaev I.P. et al., 2009; Solonenko, 2009).

The simplest equations of one-dimensional motion in a plasma flow, which describes the change in the velocity U_p and temperature T_p, of a single particle, have the following form (1, 2):

$$m_p \frac{dU_p}{dt} = C_d \times S_{mid} \cdot \frac{\rho(U_f - U_p)\left|U_f - U_p\right|}{2} \qquad (14.1)$$

$$c_p m_p \frac{dT_p}{dt} = \alpha \cdot S_{surf} \times (T_f - T_p) \qquad (14.2)$$

In the aforementioned notation: U_f and T_f are the local velocity and temperature of the plasma flow; ρ is the density of the gas (plasma) at the flow temperature T_f; S_{mid} is the midsection area of the particle; S_{surf} is the surface area of a spherical particle. The coefficient of frontal gas-dynamic resistance of the sphere C_d and the heat transfer coefficient α are calculated from empirical dependencies, which are mainly obtained in the study of low-temperature flows around bodies.

If at the initial moment of time the particle velocity is equal to zero, and the temperature is equal to the initial value T_{p0}, then in the one-dimensional approximation the velocity and temperature of a single spherical particle with a diameter D_p moving in a uniform plasma flow is determined by solution (3, 4):

$$U_p = U_f \left(1 - e^{-\frac{t}{\tau_D}}\right) \qquad (14.3)$$

$$T_p = T_f - (T_f - T_{p0})\left(1 - e^{-\frac{t}{\tau_T}}\right), \qquad (14.4)$$

In this case, m_p, C_d, ρ, S_{mid}, S_{surf}, T_{p0}, and D_p can be considered a priori known parameters in Equations (14.1) and (14.2), while the dynamic parameters of Equations (14.3) and (14.4) can be experimentally determined using high-speed video recording—constants acceleration and heating time: τ_D and τ_T. These constants have the physical meaning of the time interval that would be needed for the particle to reach the velocity (temperature) of the plasma if it moved with the current acceleration (heated up with the current intensity). Note that within the framework of this model, it is also possible to state and correctly solve the inverse problem, i.e., determination of plasma temperature and velocity based on the results of high-speed registration of tracks of calibrated particles, for example, "nano-markers" with known thermophysical properties.

14.3 METHOD FOR MEASURING THE TEMPERATURE OF PARTICLES MOVING IN A STREAM

The brightness of the measurement object is recorded by the OES based on a multi-element matrix photodetector, on a part of the photosensitive cells of which an image of a reference lamp filament is projected, the filament current of which in the calibration mode is changed according to a linear law. The values of the filament current are sequentially numbered and stored at the moments of increment by a given value of the output signal of the photodetector. The temperature of the object is determined in the absence of the filament current of the reference lamp by the stored value of the filament current corresponding to the gradation of the current output signal of the OES. The experimental stand and the spectral sensitivity of the optoelectronic path, due to the sensitivity of the C-20 photocathode and the spectral distribution of the radiation of the ZnS ZnSc Cu phosphor, are shown in Figures 14.1a, b, respectively. The spectral sensitivity (Figure 14.1b) of the optoelectronic path is due to the sensitivity of the S-20 photocathode; spectral distribution of radiation of the ZnS·ZnSc·Cu phosphor image intensifier tube; sensitivity of CMDP MT9M413.

Therefore, the OES calibration was carried out according to the TRU1100–2350 reference temperature lamp in the thermodynamic temperature range of 1800–2500 K at an effective wavelength of $\lambda = 503$ nm. The PSH-1036 power supply is turned on

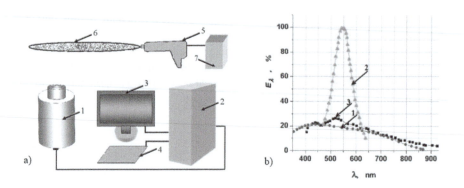

FIGURE 14.1 a) Experimental stand; b) Spectral sensitivity: 1 – S-20 photocathode; 2 – spectral distribution of the radiation of the phosphor ZnS·ZnSc·Cu image intensifier tube; 3 – **CMOS** MT9M413.

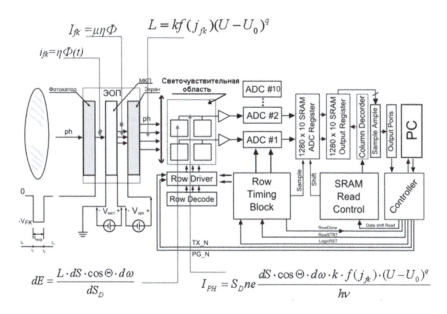

FIGURE 14.2 Functional scheme of the optoelectronic system.

in the mode of stabilizing the current flowing through the tungsten tape, which is the filament of the lamp. The initial current corresponds to the lower limit of the operating temperature range of the lamp. The experimental technique for detecting tracks of self-luminous heated particles in plasma is based on the use of specialized high-speed video cameras with parallel signal reading (Gulyaev P.Yu. & Gulyaev I.P., 2009; Gulyaev P.Yu. et al., 2012). To improve the accuracy of measuring the time-of-flight data on the motion of particles, an optical shutter with a nanosecond speed "NanoGate" was used, and to increase the sensitivity at short exposure times, a photomultiplier on microchannel plates was used, the functional diagram of the OES is shown in Figure 14.2.

Thus, the brightness of the measured object is compared with the brightness of the reference lamp, using its digital equivalent in the form of a spreadsheet of brightness and the corresponding filament currents, implemented on the basis of a macro for the ImageJ program, which improves accuracy by reducing the effect of aging of the reference lamp and prolongs its lifetime.

The OES under consideration is designed to control the temperature and speed of processes, the time of which is comparable to the time of accumulation of one frame. During the exposure, the process has time to start and end. The radiation emitted in this case is converted, reaches the screen of the image intensifier tube, and causes the phosphor to flare up. Calibration plots of the dependence of the digital signal amplitude on the brightness temperature T of the reference lamp, the allowable spread of values is ± 1 gradation, which is an error of 5% in the nanosecond and 2% in the microsecond ranges. The scatter of the digital signal amplitude is due to significant OES noise (geometric = 0, temporal = 0, ADC = ±1 gradation, residual parasitic afterglow of the phosphor ± 1 gradation) and the methodological error of contouring. In this case, the calibration was carried out on a fixed blackbody.

Micropyrometry of fast processes implies the measurement of the temperature of objects moving at high speeds. In this case, the use of calibration according to a fixed standard becomes illegal.

The method for determining the temperature of particles of the condensed phase moving in heterogeneous flows at speeds of 100–1140 m/s and having a temperature of at least 1500 K, which consists in summing the video signal along the length of the recorded track, makes it possible to use the OES calibration by a stationary black body when it operates in the multiple exposure mode. The essence of the technique is as follows.

All heated particles of the condensed phase have a continuous spectrum of their own thermal radiation. When determining the temperature of the particles, it is necessary to take into account their linear dimensions. For this, the assumption is made that all particles have a spherical shape of radius r, l are the length of the particle track.

If the particle is moving and the camera is operating in the multiple exposure mode, then the total exposure time is equal to the multiple exposure period multiplied by the number of pulses N. The total radiation flux of the particle during one operation of the image intensifier tube, i.e., during the exposure (accumulation) (5):

$$\Delta\Phi = \int_0^{t_9} \Phi(t)dt \qquad (14.5)$$

During this time, the incident radiation on the surface of the photomatrix creates an image of the particle (Figure 14.3). The basis is selected from the convenience of physical implementation, the simplicity of calculating the coefficients, and the accuracy of the approximation. The input optical image signal E (x, y), is generally a two-dimensional continuous function of continuous spatial arguments (x, y coordinates), is converted into an electrical signal described by an I(x, y)—two-dimensional continuous function of discrete spatial arguments x, y, g(x, y) is the weight function.

If the particle were motionless, then charge accumulation in the photosensitive cell would occur throughout the exposure time, illuminating the same cells. Let the

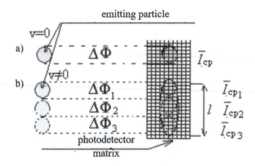

FIGURE 14.3   Discretization of the signal of a moving particle.

particle in the image have dimensions n × m. We also consider the radiation of the particle to be uniform over its entire surface. As it moves, the radiation illuminates an area with dimensions n × m. On the track length l, there are l/m of such particles. Therefore, the discretization of the signal of a moving particle (6):

$$I = \frac{L}{D} \times \overline{I} = \kappa \cdot \overline{I} \qquad (14.6)$$

If the total brightness of the track pixels at a given exposure does not exceed 255, then according to the graph of the correspondence between the brightness temperature of the lamp and the brightness gradations of the image pixels, you can determine the temperature of the moving particle. The brightness of a glowing object is less at a short exposure than the brightness of the same object at a long exposure. At different accumulation times, the particle has time to fly different distances, and, consequently, a different number of pixels will make up the track.

1. We measure the average brightness (grayscale) in the selected track obtained by video filming on the accumulation t.
2. We measure the transverse and longitudinal size of the track (in pixels).
3. We take the transverse size of its clear image as the diameter of the moving emitter. We accept that the track length contains a minimum of L = 2D pixels.
4. According to formula (14.7), we calculate the accumulation time, corresponding to the passage of the emitter in the field of view of the pixel.
5. Based on the calibration curve, we find the brightness temperature corresponding to the measured average brightness of the image on the accumulation t_1.

$$t_1 = t \times \frac{D-d}{L-D} \qquad (14.7)$$

Thus, the method for determining the brightness temperature of a moving emitter is reduced to comparing the average brightness of the track image recorded at the exposure τ and the average brightness of the image of the calibration area of the stationary reference blackbody, registered at $\tau_1 = k \times \tau$ (exposure time multiplied by the correction factor), k < 1.

14.4   BUILDING A HEATING MODEL AND ACCELERATION OF INDIVIDUAL PARTICLES

Tracking the temperature and velocity of a particular particle in a stream is possible due to the use of high-speed optical shutters and high-resolution television CMOS matrices, which significantly expands the possibilities of research in this area. In the mode of m-fold exposure (multi-exposure) with a given interval, the thermal radiation of a flying particle is fixed on one frame m times. This makes it possible to carry out a detailed analysis not only of the dynamic parameter of motion acceleration τ_D, but also of heating τ_T.

Let the particle be round, diameter d and l the length of the particle track. The total radiation of the particle during the exposure time can be characterized by the total brightness of the pixels along the track (14.8):

$$I_\Omega = \frac{4}{\pi d^2} \int I(x,y) \tag{14.8}$$

The particle track image can be represented by some integer number of pixels containing almost 100% of the signal brightness. If the particle is moving, then N = l/d particles fit on the length of the track l. Thus, the total brightness of radiation in the discrete representation will be expressed by formula (14.9):

$$I_\Omega = N \times \bar{I} = \frac{l}{d} \times \frac{1}{n \times m} \sum_{\substack{i=i_0, \\ j=j_0}}^{n,m} I_{ij} \tag{14.9}$$

where n is the number of pixels along the track along the X axis, m is the number of pixels along the track along the Y axis.

If the total brightness of the track pixels at a given exposure does not exceed 255, then according to the graph of the correspondence of the brightness temperature of the lamp to the brightness gradations of the image pixels, you can determine the temperature of a moving particle recorded by the OES in the multiple exposure mode. Stroboscopic effect when registering heated particles in the multiple exposure mode (Figure 14.4).

To determine the velocity by the time-of-flight method, it is necessary to know the boundaries of particle tracks as accurately as possible. Therefore, the video frame was previously subjected to significant processing: the dust content of the jet was removed, the track boundaries were distinguished (Isaeva & Boronenko, 2020). According to the selected tracks of particles, their speed and temperature corresponding to each of the segments are determined. Based on the obtained values, graphs of the velocity and temperature of particles versus time are plotted. Figure 14.5 shows

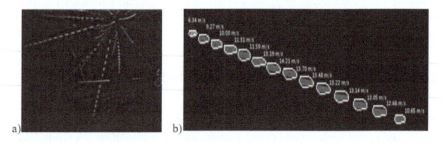

FIGURE 14.4 Stroboscopic effect when registering heated particles in the multiple exposure mode: a) tracks of random particles b) track of an individual particle.

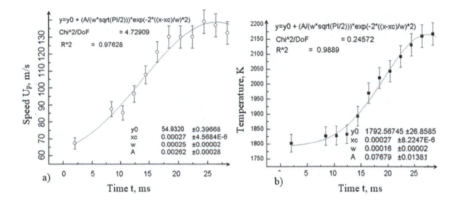

FIGURE 14.5 Determination of the acceleration time (a) and heating time (b) of particles along its track.

the results of processing the stroboscopic track of an individual particle with the approximation of the dynamic parameters τ_D and τ_T in the form of exponential solutions (14.3) and (14.4).

In plasma-arc spraying (Kharlamov et al., 2015), there can be three types of flow and disintegration of a jet of molten wire metal: metal jet flow, the formation of droplets that are almost identical in size, and, finally, the formation of droplets that differ significantly in size. The formation of droplets (900 µm or more), which differ significantly in size, can be observed with an incorrectly selected wire feed speed. A plasma torch with interelectrode inserts with a nominal power of 50 kW (MEV 50) was used to study the process of melt formation, disruption of large metal drops and their further crushing. A tungsten-cerium (2%) electrode, 1 mm in diameter, was placed into the plasma flow. The plasma torch channel remained unchanged in all experiments and had the following form: MEW section 7–8–9 mm, anode 9 mm, nozzle 14 mm with step extension up to 26 mm. The working (plasma-forming) and transporting gases in all cases are air.

In some cases, during large-drop metal transfer, the drop swings at the end of the electrode (Figure 14.6).

The formation of drops electrode metal at the framework models PIT is considered as decay cylindrical jets liquids. This model presents infinitely long jet liquids radius R, by surface which passes harmonic disturbance small amplitude α (14.10):

$$R_s = R_0 + \alpha e^{\omega t + ikz}, \alpha \ll R_0 \qquad (14.10)$$

The process of wire dispersion (the diameter of the shed drops, their shedding frequency, and the initial velocity) depends on the spatial distribution of the velocity and temperature of the plasma, the position of the molten end of the wire in the jet, the thickness of the liquid layer held at the end of the wire, and the velocity of the melt flow in it (Kharlamov et al., 2015, Khafizov et al., 2014).

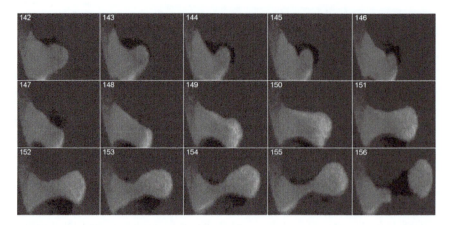

FIGURE 14.6 Formation of a drop from a liquid layer of metal at the end of a tungsten electrode (in conditional colors), 1000 fps frame sequence, 20 ns exposure

To obtain the simplest model for the formation and shedding of molten metal droplets from experimental data, it is necessary to reveal the periodicity in disturbances caused by the redistribution of heat fluxes. The thermal waves propagating in the molten electrode material appear in the image as moving light areas. The interframe difference makes it possible to establish the chronology of the appearance of perturbation regions in the heated metal and the change in their boundaries. After interframe subtraction, we represent stacks as slices, projections, or volumes in 3D-(xyz)-space (Figure 14.7a). By adjusting the viewing angle, color and contrast, you can visualize the periodicity of oscillations. Binding the centers of mass of the selected areas to specific frames of the video file allows you to build a mathematical model of disturbances in a cylindrical metal jet (Figure 14.7b, c).

The perturbation curve in the metal is well-approximated by the peak function (FWHM version of Gauss Function) (Tables 14.2 and 14.3) (14.11):

$$X(t) = R_0 + \frac{A}{w\sqrt{\dfrac{\pi}{2}}} e^{\left(-2\frac{t-t_c}{w}\right)^2} \tag{14.11}$$

The model was built without the assumption that the total current is concentrated in the volume of the molten metal jet.

When modeling the behavior of molten particles in a plasma flow, it is assumed that the coordinates of the point of entry of liquid particles into the plasma flow x0, z0 coincide with the location of the molten end of the wire. The initial values of the diameter d0 and velocity w0 of the movement of a liquid particle are determined based on the model of the jet flow of the molten wire metal and the formation of liquid metal droplets in a cocurrent high-speed gas flow (Figure 14.8).

The characteristic heating time of a particle is of the order of $\tau \sim d^2/a$, where d is its diameter and a is the thermal diffusivity of its material. If the particles in the flow are

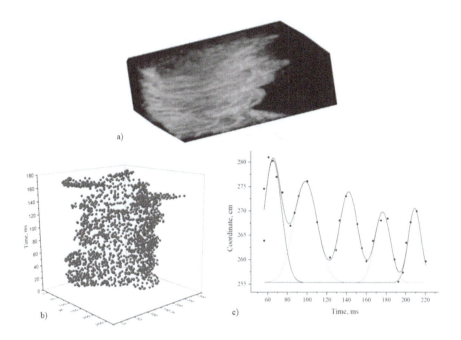

FIGURE 14.7 Visualization of the periodicity of disturbances: a) chronoscopy of disturbances on the surface of a cylindrical jet; b) instability of a current-carrying jet of heated metal during the formation of a drop; c) model of disturbances in the jet during the formation of drops.

TABLE 14.2
FWHM Version of Gauss Function (FWHM Version of Gauss Function)

Plot	Peak1(B)	Peak2(B)	Peak3(B)	Peak4(B)	Peak5(B)
y0	255.321 ± 1.363	255.321 ± 1.363	255.321 ± 1.363	255.321 ± 1.363	255.321 ± 1.363
xc	65.390 ± 1.121	98.459 ± 1.665	142.175 ± 1.265	176.323 ± 1.601	209.296 ± 1.303
w	17.597 ± 2.775	23.571 ± 4.813	18.631 ± 3.049	15.821 ± 3.468	13.065 ± 1.967
A	553.817 ± 88.944	612.711 ± 113.221	434.679 ± 71.873	287.728 ± 71.393	250 ± 0
Reduced Chi-Sqr	7.562				
R-Square (COD)	0.921				
Adj. R-Square	0.835				

single, then when filming without multiple exposures, the heating time (Figure 3.5b) and the acceleration time can still be easily determined. However, this will fail if there are many particles, because identification of particles in a dense flow is impossible.

The aforementioned experimental technique is justified in the analysis of individual particles and lightly dusty plasma jets, but with an increase in the amount of

TABLE 14.3
Summary

Peak	R₀ Value	Standard error	t_c Value	Standard error	w Value	Standard error	A Value	Standard error	sigma Value	Standard error	FWHM Value	Standard error	Height Value	Standard error
1	255,321	1,363	65,39	1,121	17,597	2,776	553,818	88,944	8,799	1,388	20,719	3,268	25,111	2,248
Peak2(B)	255,321	1,363	98,459	1,666	23,571	4,814	612,712	113,221	11,785	2,407	27,753	5,668	20,74	2,722
Peak3(B)	255,321	1,363	142,175	1,266	18,63	3,05	434,68	71,874	9,315	1,525	21,935	3,591	18,616	3,058
Peak4(B)	255,321	1,363	176,323	1,602	15,82	3,469	287,728	71,393	7,91	1,734	18,627	4,084	14,511	2,605
Peak5(B)	255,321	1,363	209,297	1,303	13,066	1,967	250	0	6,533	0,984	15,383	2,316	15,267	2,299

Reduced Chi-Sqr 7,562
Adj. R-Square 0,836

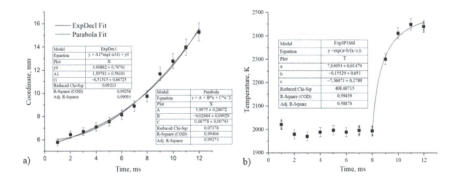

FIGURE 14.8 Video recording 1000fps, normal mode a) Determining the particle motion model (a) and particle heating time (b).

condensed phase, it is necessary to take into account its interaction with the plasma flow, which has a limited load and throughput, which is reflected in the form of a fundamental flow diagram.

14.5 OPTIMIZATION OF THE DEPOSITION PROCESS ACCORDING TO THE FUNDAMENTAL DIAGRAM

Based on the SNP method, a diagnostic complex for thermal spraying was developed (Dolmatov et al., 2016a), which contains two data acquisition channels. The first channel (based on the thermal imaging camera HD1312–1082-G2, PhotonFocus, Switzerland) collects data at a frequency of 55 frames per second, the exposure time of the camera varies within 10–1000 µs. The second channel (based on the LR1-T spectrometer, Aseq, Canada) provides registration of spectra with a frequency of up to 70 Hz and an exposure time of 2–5000 ms. In this case, if the measuring volumes of the thermal imaging camera and the spectrometer are combined, but their exposure times are made different, then this will make it possible to achieve a high signal-to-noise ratio in the measured spectrum of the ensemble of particles of the dispersed phase, and provide about 10 µs for recording tracks of fast-moving particles on images of a series of frames (Dolmatov et al., 2020).

The basis of the physical model of the collective motion of particles in plasma (Gulyaev P. Yu. & Gulyaev I.P., 2009). is the representation of a two-phase flow shown in Figure 14.9, in the form of the motion of two interpenetrating continuums of the gas phase and "pseudogas" of condensed phase particles (Boronenko et al., 2012), for which the continuity equation is valid, which describes the motion of the flow of a Newtonian fluid sensitive to shock waves (14.12):

$$\frac{\partial n}{\partial t} + \mathrm{div}\left(n \cdot \vec{v}\right) = 0 \tag{14.12}$$

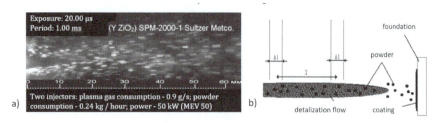

FIGURE 14.9 a) An example frame of a high-speed video filming of a two-phase stream; b) Scheme for diagnosing the velocity distribution of condensed phase particles in a plasma flow.

where n = N(t)/V is the concentration of particles in a small measuring volume V=Δl*S, and the value S is the cross section of the flow, Δl is the thickness of the measuring volume in the direction of the flow, and v is the flow velocity of the "fake gas".

The dependence of the intensity of the flow of particles crossing the cross section of the measuring volume V on the density of particles in the flow determines the required characteristic (14.13):

$$q(t,x) = Q(\rho) = \left(\frac{dn}{dt}\right) \times V \qquad (14.13)$$

where $Q(\rho)$ is the fundamental diagram of a two-phase flow.

The speed and density of particles per unit length of the flow are determined experimentally (14.14):

$$\rho(t) = \left(\frac{\partial n}{\partial x}\right) \times S = \left(\frac{\partial n}{\partial x}\right) \times \frac{V}{\Delta l} \qquad (14.14)$$

As a result, the one-dimensional continuity equation will be written in the form (14.15):

$$q(t,x) = \rho \times v \qquad (14.15)$$

Thus, it is possible to find the intensity of the flow of particles crossing the cross section of the measuring volume V by experimentally determining the average velocity of particles in the flow and their linear density, and, consequently, construct the fundamental diagram of a two-phase plasma flow shown in Figure 14.10 for the laminar plasma torch MEV 50.

The curve of the fundamental diagram reflects the change in the velocity of particles in the plasma flow at different powder loadings. The straight line at zero flux density corresponds to the plasma velocity, and the maximum of the curve marks the performance limit of the plasma torch.

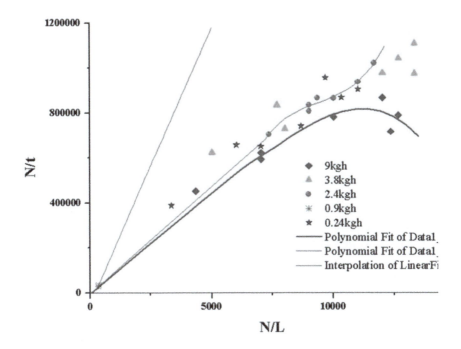

FIGURE 14.10   The curve of the fundamental diagram.

14.6   THE CONCEPT OF NETWORK CONTROL BY THE OPERATOR OF THE PARAMETERS OF THE SPRAYING PROCESS

The evolution of information-measuring systems has overcome the stages: analog device, digital device, virtual device (sensor + PC), and has reached the stage of embedded systems. The next logical step in the development of measuring instruments should be intelligent network devices, which, due to computing technologies and artificial intelligence, will not only be able to encapsulate the traffic of primitive signals, but will also gain the ability to organize into intelligent networks with a qualitatively new level of processed information (Dolmatov et al., 2021).

The basis of the network interface of an intelligent device for monitoring the parameters of the deposition process with a temporal and spatial resolution of up to 1 ms and 10 µm, respectively, was the assessment—the luminance-spectral emissivity (LSE). When building an intelligent network pyrometer, the authors used the concept of a virtual one (Dolmatov A., 2022):

- Registration of primary data is performed by a specialized hardware platform
- Processing and analysis of the received data is carried out by computing means with additional conditions

- Measuring and computing means of an intelligent network device, including support for a network interface, must be integrated into a stand-alone device
- ICP computing facilities should organize storage, as well as primary and intellectual processing of recorded data with the possibility of their synchronization with the measurement results of other ICP measuring complex
- Several ISPs can be organized into measuring complexes only on the basis of network technologies (hardware signals of the "external synchronization" type are excluded)
- Each ISP in the measuring complex can be assigned a specific role and order of user access using external administrative tools
- ISP must support a specialized networking protocol and provide tools for client software development
- The ISP should support its own Web-interface for interactive setting of individual parameters, providing access to measurement results and data processing, as well as providing an SDK to customers.

The digital spectrometer LR1-T (Aseq. Canada), operating in the range from 200 to 1100 nm with a dispersion of 0.24 nm, was chosen as the measuring platform for the pyrometer being developed. Its computing facilities are implemented on the basis of an nVidia Jetson Nano embedded microcomputer running the Linux operating system. The USB interface is used to integrate the measuring and computing platforms. nVidia Jetson Nano hardware includes 4-core ARM Cortex A57 (1600 MHz), Maxwell GPU with 128 CUD A cores. LPDDR4 is 4GB.

To implement the functions of an intelligent network pyrometer in the nVidia JetsonNano operating system, the following are deployed: MySQL DBMS—Apache http server. PHP. Python TensorFlow. The original software of the device implements the following functions:

- Exchange of service messages and data between the instrument's measuring and computing platforms (spectrometer driver).
- Storage, processing, and use of registered spectral data, temperature calibration data, and neural network parameters. When placing measurement results in the database, they are assigned a GUID generated and assigned by the ISP who received this role within the experimental complex. In addition, all recorded data receive a global time stamp, which is synchronized between all ICPs of the complex at the beginning of the experiment, and then maintained individually by each device due to the accuracy of the real-time clock. An accuracy of 50 ms is sufficient in the experiment to synchronize the ICP. The complex uses the NTP network protocol. If a smaller time error is required, then the original link-layer network protocol is used, which provides an ICP time stamp error of 10 μs in a switched network. The temperature calibration data is used by the pyrometer's internal software to correct the spectral signal, as well as to determine the brightness and spectral temperature of the object at a given wavelength. The parameters of the neural network represent its structure and the results of training in

the diagnosis of phase transitions in terms of the dynamics of conditional temperatures and CLAS.

- Network layer protocol and server for remote instrument control and data exchange.
- Tools for developing client software in C++ languages. FROM#. matlab.
- Web-based user interface for interactive setting of the exposure duration from 10 μs to 10 s, selection of the spectral range from 200 to 1100 nm, setting the speed of recording spectra, viewing spectral data stored in the database, analyzing the dynamics of conditional temperatures and CLCI, determining the temperature of probable phase transitions based on built-in AI tools, downloading software tools for developing client applications and services, displaying help information for the operation of the device and software services.
- Neural network inference for recognition of structural phase transitions in the processes of heating or cooling an object. The trained neural network, using TensorFlow and Tensor RT, uses the cores of the CUDA GPU module with a performance of 472 GFLOPS for streaming analysis of spectral data, determining conditional temperatures and object CLAS.

The work of the operator also moves to a new level:

- Calibration of the installation is still the responsibility of the operator.
- The stability of the parameters, with the steady spraying mode, is continuously monitored by automation. The correction of the spraying mode, when the parameters of the spraying process deviate from the norm (transition to an unstable mode), the operator can perform remotely.
- Data from sensors that control the main parameters of spraying, video recording of the process, etc., by the decision of the operator, go into the recording mode and send to the cloud.
- Access to the cloud storage is open to the scientific department. At the same time, external researchers who do not work for the company, but who wish to take part in the research, can request access to the video recordings from the operator. The operator confirms/prohibits the connection of external researchers (open access to files).

14.7 CONCLUSION

The study of the plasma spraying process is relevant due to its multifactorial nature and nonlinearity, complicated by the small size of objects, as well as the high speeds of their movement and heating. Stroboscopic "track" methods for measuring the particle flow velocity using high-speed optical shutters and high-resolution television CMOS matrices significantly expand the possibilities of research in this area, make it possible to build models of heating and acceleration of particles, and make it possible to determine the dynamic constants of motion and heating as individual particles in a plasma flow.

The use of the SNP method in the thermal imaging system YuNA (Yugra—Novosibirsk—Altai) makes it possible to measure the temperature of the condensed phase of gas flows with an error of less than 1% and obtain statistical distributions of particle temperature density and their velocities. Building a fundamental interaction diagram during collective motion allows you to choose the optimal spraying mode (Gulyaev I. et al., 2017). The proposed diagnostic method is recommended to be used to study the load capacity of two-phase flows, as well as an indicator of the limiting technological state of the plasma torch and the transition to unstable spraying modes. The introduction of network control of deposition parameters will allow the operator not only to remotely control the process, but also, if necessary, promptly involve third-party specialists in solving new problems or problems.

REFERENCES

Boronenko M.P. (2014). Review of the use of high-speed television measuring systems in a physical experiment. *Bulletin of the Yugorsk State University*. 2 (33). pp. 43–55.

Boronenko M.P., Gulyaev I.P., Seregin A.E. (2012). Model of movement and heating of particles in a plasma jet. *Bulletin of the Yugorsk State University*. 2 (25). pp. 7–15.

Boronenko M.P., Gulyaev P.Yu., Trifonov A.L. (2012). Determination of the fundamental flow diagram of a laminar plasma torch with a constant supply of powder. *Bulletin of the Yugorsk State University*. 2. pp. 16–20.

Craig J.E., Parker R.A., Lee D.Y., Wakeman T., Heberlein J., Guru D. (2003). Particle temperature and velocity measurements by two-wavelength streak imaging. *Thermal Spray 2003: Advancing the Science and Applying the Technology*. 2. pp. 1107–1112.

Dolmatov A.V. et al. (2016a). Automation of thermophysical studies of the process of thermal spraying of coatings. *Multi-Core Processors, Parallel Programming, FPGAs, Signal Processing Systems*. 1 (6). pp. 192–201.

Dolmatov A.V. (2020). Methods for controlling structure formation in high-temperature synthesis processes (review). *Bulletin of Yugra State University*. 2 (57). pp. 7–18.

Dolmatov A., Gulyaev P., Milyukova I. (2021). Intelligent network pyrometer for monitoring structural phase transitions in materials. *High Performance Computing Systems and Technologies*. 5 (1). pp. 172–177.

Dolmatov A. (2022). Mechatronic control system for high-temperature synthesis of materials based on intelligent measuring modules. *Yugra State University Bulletin*. 18 (2). pp. 11–21.

Dolmatov A.V., Gulyaev I.P., Gulyaev P.Y., Jordan V.I. (2016b). Control of dispersed-phase temperature in plasma flows by the spectral-brightness pyrometry method. *IOP Conference Series: Materials Science and Engineering, Tomsk*. p. 012058. DOI: 10.1088/1757-899X/110/1/012058.

Ermakov K.A., Dolmatov A.V., Gulyaev I.P. (2014). The system of optical control of the speed and temperature of particles in the technologies of thermal spraying. *Bulletin of the Yugorsk State University*. 2 (33). pp. 56–68.

Fauchais P., Vardelle M. (2010). Sensors in spray processes. *Journal of Thermal Spray Technology*. 19 (4).

Fincke J.R., Haggard D.C., Swank W.D. (2001). Particle temperature measurement in the thermal spray process. *Journal of Thermal Spray Technology*. 10 (2). pp. 255–266.

Gao F. et al. (2012). Optimization of plasma spray process using statistical methods. *Journal of Thermal Spray Technology*. 21 (1). pp. 176–186.

Gulyaev I.P. (2018a). Diagnostic system YuNA for disperse phase properties control in plasma and laser powder deposition processes. *Journal of Physics: Conference Series*. 1115 (3). p. 032072.

Gulyaev I.P. (2018b). Spectral-brightness pyrometry: Radiometric measurements of non-uniform temperature distributions. *Journal of Physics: Conference Series.* 116. pp. 1016–1025.

Gulyaev I.P., Dolmatov A.V., Gulyaev P.Yu., Boronenko M.P. (2017). *Method for spectral-brightness pyrometry of objects with non-uniform surface temperature.* Patent No. 2616937, IPC G01J 5/50: No. 2015123315: Appl. 06/17/2015. publ. April 18. p. 12. https://patents.google.com/patent/RU2616937C2/en.

Gulyaev I.P., Solonenko O.P. (2013). Hollow droplets impacting onto a solid surface. *Experiments in Fluids.* 54 (1). p. 1432.

Gulyaev I.P., Solonenko O.P., Gulyaev P.Y., Smirnov A.V. (2009). Hydrodynamic features of the impact of a hollow spherical drop on a flat surface. *Technical Physics Letters.* 33 (10). pp. 885–888.

Gulyaev P.Yu., Gulyaev I.P. (2009). Modeling of technological processes of plasma spraying of coatings of nanoscale thickness. *Control Systems and Information Technologies.* 1.1 (35). pp. 144–148.

Gulyaev P. Yu. et al. (2012). Methods for optical diagnostics of particles in high-temperature flows. *Polzunovskiy Vestnik.* 2 (1). pp. 4–7.

Hamalainen E. et al. (2000). Imaging diagnostics in thermal spraying-spraywatch system. Thermal spray: Surface engineering via applied research. *Proceedings of the International Thermal Spray Conference.* pp. 79–83.

Isaeva O., Boronenko M. (2020). Application of ImageJ program for the analysis of pupil reaction in security systems. *Journal of Physics: Conference Series.* 1519 (1). p. 012022.

Khafizov A.A. et al. (2014). Steel surface modification with plasma spraying electrothermal installation using a liquid electrode. *Journal of Physics: Conference Series, IOP Publishing.* 567 (1). p. 012026.

Kharlamov M. et al. (2015). Complex mathematical modeling of the processes of plasma-arc wire spraying of coatings. *Bulletin of Yugra State University.* 2 (37). pp. 33–41.

Kuzmin V.I. et al. (2020). Air-plasma spraying of cavitation-and hydroabrasive-resistant coatings. *Thermophysics and Aeromechanics.* 27 (2). pp. 285–294.

Li J.F. et al. (2005). Optimizing the plasma spray process parameters of yttria stabilized zirconia coatings using a uniform design of experiments. *Journal of Materials Processing Technology.* 160 (1). pp. 34–42.

Liu T. et al. (2013). Plasma spray process operating parameters optimization based on artificial intelligence. *Plasma Chemistry and Plasma Processing.* 33 (5). pp. 1025–1041.

Mauer G., Vaßen R., Stöver D. (2007). Comparison and applications of DPV-2000 and Accuraspray-g3 diagnostic systems. *Journal of Thermal Spray Technology.* 16 (3). pp. 414–424.

Mauer G., Vaßen R., Stöver D. (2011). Plasma and particle temperature measurements in thermal spray: Approaches and applications. *Journal of Thermal Spray Technology.* 20 (3). pp. 391–406.

Chen, Shuang Hchuan, et al. (2014). Structural Properties, Modeling and Optimization of Tribological Behaviors of Plasma Sprayed Ceramic Coatings. *Applied Mechanics and Materials.* 610. pp. 984–992.

Okovity V. A., Panteleenko A. F. (2015). Optimization of the spraying process of wear-resistant coatings based on multifunctional oxide ceramics. *Processing of Metals: Technology, Equipment, Tools.* 2 (67). pp. 46–54.

Salhi Z., Gougeon P., Klein D., Coddet C. (2005). Influence of plasma light scattered by in-flight particle on the measured temperature by high speed pyrometry. *Infrared Physics and Technology.* 46 (5). pp. 394–399.

Sampath S., Srinivasan V., Valarezo A., Vaidya A., Streibl T. (2009). Sensing, control, and in situ measurement of coating properties: An integrated approach towards establishing process-property correlations. *Journal of Thermal Spray Technology.* 18. pp. 243–255.

Solonenko O.P., Blednov V.A., Iordan V.I. (2011a). Computer design of gas-thermal coatings from metal powders. *Thermal Physics and Aeromechanics.* 18 (2). pp. 265–283.

Solonenko O.P., Gulyaev I.P. (2009). Nonstationary convective in a drop of melt mixing bypassed by plasma flow. *Technical Physics Letters*. 35 (8). pp. 777–780.

Solonenko O.P., Gulyaev I.P., Smirnov A.V. (2011b). Thermal plasma processes for production of hollow spherical powders: Theory and experiment. *Journal of Thermal Science and Technology*. 6 (2). pp. 219–234.

Swain B. et al. (2021). Parametric optimization of atmospheric plasma spray coating using fuzzy TOPSIS hybrid technique. *Journal of Alloys and Compounds*. 867. pp. 159074.

Swank W.D., Fincke J.R., Haggard D.C. (1995). A particle temperature sensor for monitoring and control of the thermal spray process. *Advances in Thermal Spray Science & Technology*. pp. 111–116.

Timofeev M.N., Koshuro V.A., Pichkhidze S.Y. (2021). Optimization of parameters of plasma spraying of titanium and hydroxyapatite powders. *Biomedical Engineering*. 55 (2). pp. 121–126.

Valarezo A., Choi W.B., Chi W., Gouldstone A., Sampath S. (2019). Process control and characterization of NiCr coatings by HVOF-DJ2700 system: A process map approach. *Journal of Thermal Spray Technology*. 19. pp. 852–865.

Wroblewski D., Reimann G., Tuttle M., Radgowski D., Cannamela M., Basu S.N., Gevelber M. (2010). Sensor issues and requirements for developing real-time control for plasma spray deposition. *Journal of Thermal Spray Technology*. 19 (4). pp. 723–735.

Zhang W., Sampath S.A. (2009). Universal method for representation of in-flight particle characteristics in thermal spray processes. *Journal of Thermal Spray Technology*. 18. pp. 23–34.

Index

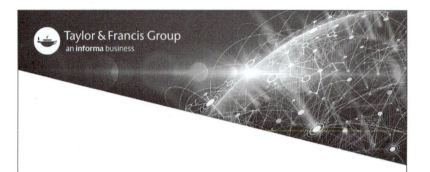